Global Engineering

Design, Decision Making, and Communication

Industrial Innovation Series

Series Editor

Adedeji B. Badiru

Department of Systems and Engineering Management
Air Force Institute of Technology (AFIT) – Dayton, Ohio

Global Engineering

Design, Decision Making, and Communication

Carlos Acosta
V. Jorge Leon
Charles Conrad
Cesar O. Malave

CRC Press
Taylor & Francis Group
Boca Raton   London   New York

CRC Press is an imprint of the
Taylor & Francis Group, an **informa** business

CRC Press
Taylor & Francis Group
6000 Broken Sound Parkway NW, Suite 300
Boca Raton, FL 33487-2742

First issued in paperback 2017

ISBN 13: 978-1-138-11446-3 (pbk)
ISBN 13: 978-1-4398-1155-9 (hbk)

Library of Congress Cataloging-in-Publication Data

Global engineering / Carlos Acosta ... [et al.].
    p. cm. -- (Industrial innovation series)
  Includes bibliographical references and index.
  ISBN 978-1-4398-1155-9 (hardcover : alk. paper)
  1. Engineering. 2. Engineering--International cooperation. 3. Engineering--Decision making. 4. International business enterprises--Management. I. Acosta, Carlos, 1954 Jun. 2- II. Title. III. Series.

TA153.G55 2010
620.0068--dc22   2009025702

Visit the Taylor & Francis Web site at
http://www.taylorandfrancis.com

and the CRC Press Web site at
http://www.crcpress.com

Contents

Section II   Case Studies: Cultural Emphasis

Section III   Case Studies: Engineering Predominance

Section IV Case Studies: Applying Concepts

Preface

Global Engineering: Design, Decision Making, and Communication is a book to be used by engineers to learn about how to better design, make decisions, and communicate in an international working environment. Today companies expect to hire engineers who are effective in a global business environment. Pressured by needs from industry and new accreditation criteria, for the past years universities have attempted to incorporate globalization as an important topic in their engineering curricula. One prevalent problem encountered by those developing such curricula at engineering programs around the world has been the lack of a framework and book specifically aimed to teach the material to engineers. This book is aimed to bridge this gap. The authors represent multiple engineering and nonengineering disciplines (industrial, mechanical and manufacturing engineering, and organizational communication). The contents of the book reflect the authors' multidisciplinary perspective as well as their experiences working in projects around the world.

The content of *Global Engineering* is based on original research involving multinational organizations operating outside of their home countries. Our research was sponsored in part by the National Science Foundation (NSF) project DMI-0116635 in the United States and its Mexican counterpart, the Consejo Nacional de Ciencia y Tecnologia (CONACyT), project 35981-U. Its goal was to develop a group of case studies that characterized and formulated how globalization affects engineering decisions that could be incorporated into engineering curricula. Although at the very beginning the research was intended to focus on companies operating in North America, very soon this had to be extended to include countries in Europe and Asia. It is not the intent of the book to take sides on the issue of globalization. The book presents globalization as a prevalent phenomenon affecting both the way companies operate and most engineering functions.

Many books have been written about globalization, but most of them are more suitable for students in the fields of business or the humanities. This book is aimed specifically at engineering undergraduate and graduate students and can be used as a text for a course on global engineering, or a reference for design courses with significant international component. The cases in the book can also be used as a supplemental text in a wide range of traditional engineering courses that would benefit from real-world case studies. To the best knowledge of the authors and the reviewers of the book manuscript, this is the first book specifically aimed at engineering students. The authors have used some of the material to teach courses at Texas A&M University and Universidad de Las Americas (Cholula, Mexico) with very positive reception by students from a variety of engineering disciplines (chemical engineering, electrical engineering, industrial engineering, mechanical engineering, and systems engineering) as well as from engineering students in organizational communication courses.

All case studies are based on real industrial projects that either the authors or their students had firsthand information and direct interviews with the characters in the case. The case study format adopted in the book allows a natural presentation of critical technical and nontechnical concepts and their complex interactions. The themes in the case studies are varied, ranging from design to supply chain and logistics problems and system improvement projects.

The book is organized in four units. Section I contains the theory, models, and decision tools that engineers need to have to formally incorporate globalization factors in their engineering work. Section II contains three international case studies where cultural factors were crucial and engineering factors were of secondary importance. Section III contains three case studies in which engineering factors were predominant, while cultural factors served as an important backdrop. The last section, Section IV, contains two case studies aimed at applying the concepts learned in the book. At the end of each chapter there are review and study questions: (1) exercises where students can practice how to use the global engineering model to analyze the case study, (2) exercises where students can practice how to use a formal tool to make decisions based on the case study facts, (3) specific technical problems associated with the chapter topics, and (4) reflection and open-ended questions. Fully worked solutions are available in the solutions manual (www.crcpress.com/product/isbn/9781439811559).

All instructors adopting the book or parts of the book are encouraged to cover Section I (i.e., Chapters 1, 2, and 3). Chapter 1 introduces globalization, Chapter 2 presents a model for global engineering, and Chapter 3 presents a formal decision-making methodology useful in situations where the engineer must consider quantitative and qualitative information in making a decision. In addition to Section I, the instructor should include at least one case study from Section II (when culture is crucial) and one from Section III (when engineering is crucial). If globalization is taught as a one-semester course, then the book can be used as a text for this course. When globalization is taught as a supporting topic as part of another course, the book can be used as a source for team projects. The instructor can ask all students to read Section I, and then assign a particular case for each team to study, prepare a class presentation, and write a project report. If multiple case studies are assigned to a team, it is recommended that they are selected from different units in the book.

It is our hope that the book will be a useful resource for engineers. This book would not have been possible without the support from the companies that provided access to information and personnel for interviews and presentations. More importantly, they guided us in the determination of what aspects of globalization are important for global-ready engineers. We are grateful to several friends in industry, colleagues in academia, and students who were instrumental in the preparation of this book, in no particular order: Michael Sabados, Rene Villalobos, Adoracion Rodriguez, Huiyan Zhang, Martin Bodewig, Haydee Serna, Kumsun Kijtawesataporn, Luis Mar, Ricardo Coronel, Roberto Gonzalez, Rafael Sanchez, Ruben Alcantara, Alejandro Garcia-Blazquez, Alejandro Lopez, and many others that cannot be named for confidentiality reasons.

Carlos E. Acosta
V. Jorge Leon
Charles Conrad
Cesar O. Malave

Section I

Theory, Models, Decision Tools

1

Engineering in a Global Age

A fundamental shift is occurring in the world economy. We are moving progressively further away from a world in which national economies were relatively isolated from each other by barriers to cross-border trade and investment; by distance, time zones, and language; and by national differences in government regulation, culture, and business systems. And we are moving towards a world in which national economies are merging into an interdependent global economic system, commonly referred to as globalization.

Charles W. Hill[1]

The strongest evidence of globalization is the increase in trade and movement of capital during the latter half of the twentieth century. Between 1950 and 2001, world exports increased by twenty times, and the rate of increase accelerated after the end of the Cold War. The main drivers of globalization are the declining trade and investment barriers as well as advances in communication, transportation, and information technologies. The costs of communication have fallen dramatically. Cell phones, fax, and Internet unite people throughout the world for pennies or fractions of pennies. Information technology has had similar effects on the speed and cost of processing global business orders. Fast, worldwide transportation is available by either airplane or containerships, and the percentage that transportation costs play in the pricing of a product has declined. As a result, many daily items used by consumers around the world are produced in distant locales—our clothes, food, and cars.[2]

When combined with a historical trend toward market economies, these technological changes have made it possible for companies to function using the best resources no matter where in the world. The resources available are of a wide variety, including money, state-of-the-art technologies, know-how and scientific discoveries, raw materials, components, and human resources. As a consequence, companies that lack a global perspective and access to the full range of globally widespread resources may not be competitive enough to survive and succeed in today's business environments.

Globalization also imposes new demands on engineers and engineering organizations. With real-time Internet communication and worldwide logistic networks, it seems easy and efficient to produce and distribute the same products throughout the global village.[3] Initially, scholars studying globalization argued that the tastes and preferences of customers would converge into one global norm, helping to create a global market and simplify product design and development.[4] In some cases these predictions have been confirmed. Companies like McDonald's or Coca-Cola have succeeded in gaining customer acceptance of their worldwide standardized products on a global basis. However, many other producers have found that customer tastes and preferences have forced them to design and produce for specific markets, each of which contains distinctive circumstances and cultures.[5] For instance, this phenomenon can be observed at McDonald's, where menus include special items designed for the taste of the locals, depending on the country in which they operate. An extensive case study in the automotive industry is presented in Chapter 6. Localization is the process engineers use to adapt designs to meet the language, cultural, and technical requirements specific to the market's region to be able to offer a competitive product in each market. In this process of localization, engineers must consider many region-specific influence factors. Production is also affected by different circumstances of each country. Legislation, like environmental protection laws; economic conditions, like the labor costs; or technological factors, like the availability of needed goods have to be considered for localization. Therefore, not only products, but also production processes can be localized. Particularly in technology transfer projects, processes have to be adapted to local needs. Figure 1.1 sketches several localization influence factors.

As markets globalize, the need for standardization in organization, product design, information systems, and manufacturing processes increases. Yet managers are also under pressure to adapt their organization to the local characteristics of the market, the legislation, the fiscal regime, the sociopolitical system, and the cultural system. This balance between consistency and adaptation is essential for corporate success (Trompenaars, 1998). To be able to offer products at competitive prices, a global production network is needed. Standardization of products is a way to reach high

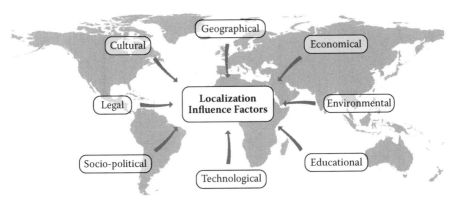

FIGURE 1.1
Localization influence factors.

production volumes and lower costs per unit, exploiting the economies of scale. Especially cars with their high development costs have to be manufactured in high volumes. A single carmaker's factory is then responsible to manufacture a specific car model for all target markets. But, adaptation is expensive because it sacrifices the economic advantages of specialization and economies of scale.

Two concepts have been proposed to help organizations deal with these dilemmas. The first is *glocalization*. Products are first globalized (standardized) and then have to be localized (adapted) to local conditions. The neologism *glocalization* was formed through the combination of the two terms *globalization* and *localization* (Fitri, 2004). It represents the creation of products for the global market, but is adapted to suit local conditions. The second concept is *homologation* (from the verb *homologate*, meaning "to approve or confirm officially"), the certification of a product or specification to indicate that it meets regulatory standards. Homologation departments are concerned with achieving regulatory compliance. In case of a foreign target market, homologation includes the product certification, with or without adaptation, to local legislation. It is a very important part of localization, but is a narrower concept because it does not encompass other important adaptations, e.g., to fit cultural values and practices.

Consider the following scenarios:

> Biomedex (a fictitious name) is a medium-sized Belgian pharmaceutical firm. It owns a laboratory and production facilities scattered all over the country. Biomedex began operating its first manufacturing site in Eastern Europe and exploratory biotechnology research in the early 1970s.... Production of a commercially viable biotechnology product would take several years. To proceed further, Biomedex would need to raise large sums of capital to find an extensive research and product development phase.... Unfortunately, Belgium's stock market was small, lacked liquidity, and was segmented from international markets ... the limited liquidity and conservative nature of Belgium's capital market would have made it very costly for Biomedex to raise capital in its own country. Faced with dilemma, Biomedex contacted Morgan Grenfell, a London-based commercial bank.... Morgan Grenfell arranged for Biomedex to list its shares on the London Stock Exchange to facilitate conversion and gain visibility in 1979.... At this time, biotechnology was attracting the interest of the U.S. investment community.... Biomedex then decided to explore the potential for a U.S. stock offering and listing of its shares on the New York Stock Exchange (NYSE).... As it turned out, ... raising an additional $50 million for Biomedex. The funds had been spent for conducting R&D activities and establishing another foreign production facility in Neighbor country.
>
> Coca-Cola is the largest selling soft drink in the world but sales vary by nation. The average American consumes almost 291 twelve-ounce servings of Coke annually.... The average annual number in German, Spain, Belgium, and Austria is 166; in Great Britain, Ireland, and Switzerland

it is about 110; France, Italy, Portugal lag behind with consumption at an average of 65. However all of this is in the process of changing. Coke has undertaken a vigorous campaign to dramatically increase consumption in Europe. One of the first strategic steps has been to replace local franchisers over to more active, market-driven sellers. In France, Pernod, a Coca-Cola franchisee, was forced to resell some of its operations back to Coke and a new marketing manager was appointed.... As a result, per capita consumption in France has increased by 65%.... In England, ... new marketer began to run contest and sponsor event all over the country ... resulting in triple sales.

... Mercedes has touched down in rural, rough-hewn Vance, Ala. (population 400). It's about the last place on earth you'd expect to find the buttoned-down German automaker, but Mercedes has managed to negotiate the various speed bumps and slick patches that come with setting up a [new] ... factory in a foreign land. Production on a new M-class sport-utility vehicle, priced around $35,000, began at the Vance plant earlier in 1997.... Mercedes had some solid reasons for setting up a plant in the U.S. Back in Germany labor costs are about 50% higher than in the small-town American South.... Top Mercedes honchos settled on tiny Vance after considering 150 sites in 30 different states, including Nebraska and such usual Sunbelt suspects as North and South Carolina. To help its case, Alabama pledged a whopping $250 million in tax abatements and other incentives.... It's not certain how Mercedes' M-class will ultimately fare in the marketplace against such established competition as the Jeep Grand Cherokee and the Ford Explorer. Yet the company is clearly gaining valuable experience in how to set up and operate a plant in a distant land, and that experience will be needed soon. It just so happens that Mercedes will start producing its A-class sedan in a factory in Sao Paulo in 1998. So, in the not so distant future, Mercedes may well transfer an employee born in deepest Alabama to its operation in sultry Brazil....[6]

As illustrated above, these organizations "go global" for a number of different reasons. Biomedex was seeking financial resources outside its home country that could support the growth and competitiveness of its business. Coca-Cola was interested in increasing its presence in the European market. Mercedes-Benz, like its German peer BMW, realized that it could substantially reduce its labor costs by almost 50% and move closer to its U.S. market by locating in the southern United States. Table 1.1 shows just how different compensation is across different countries.

In addition, firms usually receive other indirect benefits, such as tremendous tax abatements from host governments and the invaluable experience of operating an offshore facility in a major market.

The job of an engineer has been deeply affected by this global transformation of companies. Engineers working in research and development, design, production, service, and other areas can be located anywhere in the world as required by the business. This includes national engineers located abroad,

TABLE 1.1

Hourly Compensation Costs in U.S. Dollars for
Production Workers in Manufacturing

Country	1990	1995	2000
France	15.49	19.35	15.66
Germany	21.81	31.60	24.01
Japan	12.80	23.82	22.00
Korea	3.71	7.29	8.48
Mexico	1.58	1.65	2.08
Portugal	3.77	5.37	4.75
Sri Lanka	0.35	0.48	0.48
Taiwan	3.90	5.85	5.85
UK	12.70	13.78	16.45
United States	14.91	17.19	19.72

Source: Data from Bureau of Labor Statistics, 2002.

or foreign engineers working for the company from their country of origin. Globalization has accelerated and broadened the outsourcing of business functions to include international contractors, suppliers, and services. The exacerbation of business specialization and outsourcing requires a change in how engineers approach, model, and formulate problems. Instead of having a narrow perspective on specific products produced in a single facility, they must now view the company as globally dispersed enterprises with materials, information, and people routinely moving across organizational and national boundaries. Also, the designs must be versatile to facilitate the variations needed to adapt the product for performance and acceptance in heterogeneous market conditions. To make things more challenging, not only the job market but also the engineering resource pool has become international and competitive, with more and more well-qualified engineers graduating from universities around the world.

To succeed as an engineer in this global environment, the core technical knowledge required by engineers must be augmented by nontechnical knowledge, including business, interpersonal communications, culture, geography, international laws and trade regulations, and other "soft" skills. To have an edge in the job market without additional soft skills, a more technically oriented engineer will have to have advanced knowledge in his or her field and excel at creativity and innovation. However, increasing trends in the production of engineering researchers in developing countries suggest that global competition in advanced technical skills is already playing a role and will become routine in the near future. Engineers without these soft skills, especially those in advanced countries, will find themselves out of competition for jobs against equally qualified engineers from other countries willing to work for a fraction of the compensation.

Engineeering, Change, and Resistance

In important ways, every case study we examine in this book involves change. Engineers create and implement new technologies, or new ways of using traditional technologies. Consequently, change is built into the nature of engineering. However, with change comes uncertainty and insecurity, and with insecurity comes resistance. When changes are unexpected, or upset comfortable ways of doing things, or force people to change their interpersonal relationships, that uncertainty is magnified, and the incentives to resist change are increased. Consequently, resistance is most likely in high uncertainty avoidance (UA) cultures, but some degree of resistance is inevitable whenever change occurs.

Culture influences resistance in a second way. In all societies people learn distinctive ways of managing uncertainty. Some rely heavily on rules, others on interpersonal relationships, and still others on institutions like the family or an organization. If a particular change upsets the coping mechanisms of a culture, resistance is even more likely. For example, new technologies often are implemented in ways that change organizational structures. Some employees may be fired, others transferred to a different division, and others suddenly have to deal with a supervisor they have never met before. When the change involves an organization from one culture taking over a plant or company in another culture, these changes are even more likely. People in the home office feel most comfortable and most secure if their employees are in charge of the new operation, and if the company quickly installs its normal way of doing things in the new locale. But, doing so maximizes the changes that the offshore employees experience. If they have come to expect lifetime employment with the same firm and value loyalty between workers and organizations, layoffs will be very threatening. If they feel secure because the company has always hired family members, and promoted people based on their connections to the organization, bringing in new employees because of their technical expertise or connections to the home office will create a great deal of insecurity and mistrust. If the employees of a firm have a great deal of pride in their company and the way it operates, change will be both disconcerting and insulting.[7]

Resistance also is more likely if change is mismanaged. The most common source of failure in change efforts is failed communication. If the employees who will be affected by the change are involved in planning for it from the beginning, if the company is open and honest about the reasons for the change, if management persuades employees that the organization really does need to change and that their lives will be better or more stable as a result of the change, the likelihood of resistance is reduced.[8] If the change quickly begins to yield results that employees perceive are positive, resistance can quickly dissipate. However, if employees' expectations about the advantages of the change are violated, either because management promised more than the change can deliver or because it fails, resistance will escalate.[9] Unfortunately, the agents of change may not even know that resistance is occurring. Overt resistance is dangerous, so most resistance occurs quietly. The simplest version is for employees to support the change in public, but not change their behaviors. Covert resistance may be more active, as employees find creative ways to undermine the change. The history of pay-for-performance systems—in which local managers continue to make salary and promotion decisions based on the processes traditional in their cultures and then create rationalizations of their decisions that are consistent with the desires of the home office—provides an excellent example of passive resistance in operation.[10] Ironically, passive resistance may actually assist change over the long term. Employees rarely resist every part of a planned change. Over time they find ways to live with some of the changes, make adjustments in other aspects of the change that make it more palatable, and become accustomed to other aspects. Although the resulting change may not be exactly what management envisioned or the engineers responsible for implementing a change wanted, it may come to be accepted, and may even be a better system than the planned change. Time increases familiarity, and reduces the uncertainty caused by change. In addition, change agents may realize that the process is not going well, successfully diagnose the reasons for its failure or for employee resistance, and make appropriate changes. As a result, over the long term organizational change efforts may be more successful than they had a right to be, given the ways in which they were implemented initially.

Conclusion

The twenty-first century brings a challenging new world to professional engineers and engineering students. As always, technical expertise, operational skill, and creative problem solving will be absolutely necessary for a successful career. But, increasingly those traditional abilities will be necessary but not sufficient. Understanding the situations that surround the development and implementation of new technologies, from organizational processes to cultural guidelines and constraints, also will be crucial. In the remaining chapters of this book we will present case studies of global engineering. In each case, cultural factors, and intercultural communication, are important aspects of technical change. But, cultural factors are more important in some cases than in others. Similarly, new technologies and the complications they create sometimes are critically important, sometimes less so. Consequently, we have arrayed our case studies on a continuum, beginning with those in which culture is crucial, and technology less central to understanding the case, and moving progressively toward those cases in which technology dominated culture. Our final chapter is a test of sorts. We report on a successful change effort, but we are not certain about why it succeeded. We describe the "facts" of the case and suggest some plausible explanations, but we leave it up to our readers to complete the case analysis. Hopefully this exercise will allow readers to check on their understanding of the previous case studies, and to get a sense of just how complicated technological change can be.

Review and Study Questions

1. Select a manufacturing company of interest to you that has global operations and identify the main drivers for them to have a global presence.

2. What were the five largest economies in the world 10 years ago? Compare the list to the five largest economies today. Has globalization been an important factor for the differences? Use gross national product and gross national product per capita to measure the size of the economies.

3. China has been viewed as an "infinite source of low-cost labor" for the past decade. Do you think this will continue to be the case in the next decade? Formulate two arguments supporting your answer.

4. Are the concepts of globalization and localization contradictory or complementary? Can they be compared?

5. Give an example of how a product design has been altered due to localization.

Notes

1. Charles W. Hill, *International Business: Competing in the Global Marketplace* (Burr Ridge, IL: Irwin, 1994).
2. Ibid.
3. Theodore Levitt, "The Globalization of Markets," *Harvard Business Review* 61 (May–June 1983): 92–102.
4. Ibid.
5. Alejandro García, Research on the Technical and Cultural Factors That Affect the Implementation of Total Productive Maintenance Standards on New Technology Transfer Projects in the Tlaxcala Region, Thesis, Universidad de las Américas-Puebla, Mexico, 2003.

6. Hill, *International Business*, p. 315; A. M. Rugman and R. M. Hodgetts, *International Business: A Strategy Management Approach* (New York: McGraw-Hill, 2000), p. 4; Justin Martin, "Mercedes: Made in Alabama," *Fortune* (July 7, 1997), information available at http://www.fortune.com/fortune/articles/0,15114,380011,00.html.

7. Ronald Havelock, *Planning for Innovation* (Ann Arbor: Center for Research on Utilization of Scientific Knowledge, University of Michigan, 1970). For classic studies of resistance to change, see David Collinson, *Managing the Shop Floor* (New York: DeGruyter, 1992); Donald Schon, *Technology and Change* (New York: Dell, 1967); James Scott, *Domination and the Arts of Resistance* (New Haven, CT: Yale University Press, 1990); and Gerald Zaltman, Robert Duncan, and Jonny Holbek, *Innovations and Organizations* (New York: John Wiley, 1973).

8. K. W. Deutsch, *The Nerves of Government: Models of Political Communication and Control* (New York: The Free Press, 1963); Ronald Pellegrin, *An Analysis of Sources and Processes of Innovation in Education* (Eugene, OR: Center for the Advanced Study of Educational Administration, 1966); Schon, *Technology and Change*. For a comparative analysis of communication in successful and unsuccessful organizational change, see Charles Conrad, *Strategic Organizational Communication*, 3rd ed. (Ft. Worth, TX: Harcourt Brace, 1994), pp. 255–57.

9. Rodney Coe and Elizabeth Barnhill, "Social Dimensions of Failure in Innovation," *Human Organization* 26 (1967): 149–56.

10. For extended analyses of passive resistance, see Chris Argyris, *Intervention Theory and Method* (Reading, MA: Addison-Wesley, 1970) and Anthony Graziano, "Clinical Innovation and the Mental Health Power Structure," *American Psychologist* 24 (1969): 10–18.

2

A Global Engineering Model

The global engineering model (GEM) described in this chapter is aimed at helping engineers think with a global perspective. The model (Figure 2.1) defines three main dimensions of globalization that can be used by engineers to make sure that all important global aspects of a project are considered, or to formally compare global engineering projects.

GEM is constituted by three layers: the global orientation layer, the operational layer, and the context layer. These layers are interdependent and represent the human, business, and socioeconomical aspects, respectively, of an engineering project.

The core of GEM is the global orientation layer. This is the people's layer and is to a great extent the product of its history. Global orientations are influenced by the norms and institutional arrangements of the organization's home culture. For example, when U.S. firms grow, they usually do so by merging with or acquiring other firms. The U.S. political, legal, and economic system is designed to facilitate growth of this kind. Members of those institutions have learned to view mergers and acquisitions as the normal way of expanding. As a result, when U.S. firms move overseas they tend to do so by merging with or acquiring local firms or facilities, in spite of abundant evidence indicating that this approach *maximizes* cross-cultural complications. History and culture combine to create very stable global mindsets.

The second ring in the model includes the conditions that may lead an organization to go global—improving the efficiency of its organization, operations, and supply chain, enhancing market participation, and obtaining needed resources. Firms that have succeeded in particular kinds of global operations in the past tend to develop structures and systems that facilitate future moves; firms that have failed not only become reticent about further global efforts, but they are not likely to develop the systems and structures that would help them succeed. Firms will be most successful and most profitable if they are able to minimize the total costs of their operations, successfully exploit the most profitable markets for their products or services, and minimize the costs of resources, including raw materials and labor. This layer involves answers to questions such as: What else does the company need to achieve global competitiveness? What factors would make the greatest contribution to the company? When will the company need to globalize those factors? Where would a practical alternative company be placed? What are the benefits that a company can exploit from local conditions in term of

Globalization is complex and multifaceted:
A Global Engineering Model (GEM)

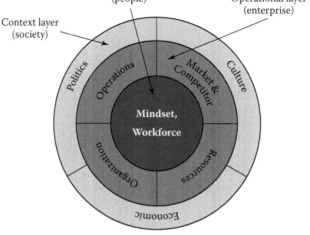

FIGURE 2.1
The global engineering model.

these three factors? This layer also includes potential operational barriers and sources of system breakdown.

The outermost layer of the model involves societal considerations—political, economic, and cultural factors. For example, even if a company can minimize its production and operational costs by moving its factories to another country, it may find that political instability in that country counteracts the economic advantages of relocating. Similarly, the potential for culture clash between expatriates and local workers may be so great in some situations that the financial advantages of relocating are offset.

We will explain each of these levels in more detail later in this and other chapters, but at this point it is important to understand that all of these factors influence the success of global engineering, and that they are related to one another in complicated ways. For example, an organization's mindset (level 1) may be based on past successes in obtaining inexpensive resources (level 2). Given that history, the organization is likely to think of low cost when it considers global operations. It may even have created a division whose job it is to seek out low-cost inputs. But, that mindset may lead it to overlook opportunities to build a market for its products or to improve operations and supply chains (level 2). For example, if a firm succeeds in locating a low-wage area, and succeeds in keeping wages low, its offshore workers may not have the income necessary to purchase its products. It also may create increased political opposition (level 3) to its operations, or signal its competitors that they too can minimize their production costs by relocating to the area (level

3, economics). For example, BMW built production facilities in the southern United States for the same reasons that Mercedes did, and at almost the same time. When one U.S. textile firm moved its operations to Mexico, and later to China and Southeast Asia, others followed suit. However, once competitors duplicate an organization's offshore operations, the strategic advantages of the location disappear. These effects may in turn lead the company to cling to its global mindset even more ferociously, obsessively looking for even lower cost areas for its production facilities, continuing the cycle. Or, they may lead the company to broaden its mindset, and start thinking in terms of global operations and a global market. Our point is that global engineering is a very complicated process. It involves multiple dimensions, and each of these dimensions influences the other dimensions.

The Global Organization Layer

The global orientation has two main components: the global mindset of the people working in the organization, and the global organization (Figure 2.2).

Perhaps the most important reality of the twenty-first century is that all organizations, whether international or domestic, need to be more global in their outlook.[1] This does not necessarily mean that all organizations will develop global operations. Neither does it mean that all employees and all divisions of an organization will need to adopt a global mindset with the same intensity. It does mean that key decision makers in all organizations will need to better appreciate the challenges involved in remaining competitive in the face of globalization, and that some employees, those whose jobs

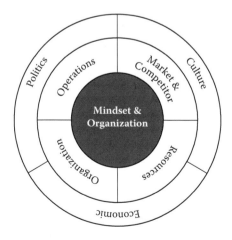

FIGURE 2.2
The global organization layer.

TABLE 2.1

Global Mindset

Mindset	Outlook	Strategy
Ethnocentrism	Centralized/controlled	International
Polycentrism	Decentralized/autonomous	Multinational
Geocentrism	Networked/interdependent	Transnational/global

require them to engage routinely in cross-border activities, will be required to seek and understand a more global mindset. Unfortunately, some employees can make this transition more easily than others. Some take an *ethnocentric perspective* (Table 2.1), using their home-based culture and experiences as reference, and attempt to replicate those successful formulas in their overseas operations.[2] In fact, some researchers believe that one of the most important elements of a culture is the extent to which people believe that their own ways of doing things are universally applicable. Of course, people with ethnocentric beliefs will find it difficult, if not impossible, to adapt their values or practices to different cultures.[3] Other people take a *polycentric perspective*. They maintain the perspective of the home office, but also include some of the host country's standards and practices in their overall operations. They also tend to be more involved personally with members of the host society, involved in more international activities, and have greater exposure to the differences between home and host countries. In other cases, organizations develop *geocentric perspectives*, in which the perceptions and decisions for all company activities will be globally based. Decisions are made through complex, international networks of employees who focus their attention on interdependence and coordination between the parent and the affiliates all over the world. A particular organization's choice of perspective depends in part on its experience in multinational operations, and partly on how open its home culture is to diverse perspectives.

Developing a geocentric mindset involves three phases: openness, awareness, and integration. It begins by opening the organization's "mind" toward diversity.[4] Openness refers to the extent to which people in the organization are open to diversity across global or multicultural differences and markets. Upon attaining a high degree of openness, an organization will be recognizing and learning differences, which will lead to the next step, awareness. Awareness refers to the extent to which people in an organization have the knowledge concerning diversity across global or multicultural differences and markets. Awareness alone will not bring an organization to a successful globalization. Integration is the ability to leverage the strengths of different groups and cultures. A firm that is aware of global differences will not have a successful global mindset unless it can convert its awareness into an integrated knowledge of how to manage multiple cultural, political, and economic differences. For example, British Oxygen Company (BOC) helped cultivate a global vision by establishing a team called Best Operation Practice.

This global team travels around BOC subsidiaries to witness the local operating practices. This approach allows the team to gather firsthand information directly from operators all over the world. The team then compiles the best practices that it discovered and designs new operation standards according to the obtained information. They then distribute them as the worldwide operation procedures. They also realize that local variation may be necessary, so they enlist a local manager to tailor the procedure according to the local realities. The process is continual because the organization realizes that situations across the world are constantly changing.

The BOC example brings us to the second dimension of a global orientation, supporting structures. Mindsets are important, but they are implemented through organizational structures and practices. Organizations that adopt an ethnocentric mindset tend to create highly centralized structures to implement it. The global strategy of a company is held only in top executives' hands. Typically, implementing it involves entering a market by acquiring an existing plant or organization. Local employees are expected to adapt their behaviors to the standards of the acquiring firm. The likelihood of culture clash is magnified. Those clashes are resolved by force—the corporation's agents are placed in charge and employees are required to accept the systems, rules, and production technologies used in the company's home operations—but tend to recur.[5] Issues are pushed up the chain of command to senior management in the home office, overloading their own agendas, causing delays in decision making, and leaving them little time to focus on broader issues of leadership. In addition, centralization provides little opportunity for lower-level employees to understand, much less come to support, the organization's global mindset.[6]

A polycentric mindset typically leads to more decentralized structures and modes of decision making, especially between the firm's home office and its foreign operations. The focus shifts from a few individuals at the top of the organization being oriented toward emerging global challenges, to having everyone involved. Senior management passes the responsibility of global matters to local units, increasing their self-government and, meanwhile, encouraging each unit to develop ways to coordinate its activities with every related unit. The organization as a whole becomes more flexible, and each unit can adapt to local situations. Each unit can learn from the experiences of the other units, and that learning can be integrated into the overall design and operation of the organization.[7] However, the overall structure moves from the simple pyramid shape of a bureaucracy to a much more complicated network or matrix. The organization is able to scan the world from a broad perspective and look for unexpected trends, opportunities, or threats. Unlike the centralized, ethnocentric strategy, in which the focus is on downward communication from the home office in order to control operations in other countries, the focus is on communicating upward and lateral communication in an effort to learn as much as possible about global situations and devising ways to strategically adapt to the global context. Participation between expatriates and local colleagues will effectively cultivate a truly global mindset on both sides, and later the entire organization. This approach has been used successfully by many Japanese firms to compensate for their very bureaucratic structures and homogeneous upper management team.

However, the resulting organizational structure is much more complicated than that of the traditional bureaucratic pyramid used in more ethnocentric firms. A decentralized firm takes on a structure that resembles a matrix or a *heterarchy* in which responsibility is shared, decision making is dispersed among the various units, and employees begin to see themselves as part of an organic worldwide entity. Management's primary task becomes one of coordination rather than control, and local managers focus on finding ways to exploit similarities across countries rather than emphasize differences.[8] For example, McDonald's has a strict policy for ensuring the quality of its food. Therefore, food processing technology is centralized throughout the world. Local units use almost the same method to cook the main menu items, such as the Big Mac, french fries, etc. Any major deviation regarding this will be made from the parent company only. However, McDonald's also is aware of the great variability of local tastes, so control over special menu items is decentralized, leaving some room for its subsidiaries in any country to develop their own menu items, such as the Teriyaki burger in Japan, the McBurrito in Mexico, and three different kinds of vegetarian sandwiches in India, and products to local flavors and requirements. But, sustaining high levels of coordination in a global operation is very difficult. A 1989 study of U.S competitiveness by the Massachusetts Institute of Technology (MIT) found that one factor that hurts U.S. firms in the global economy is their failure to achieve cooperation both within a company and between companies.[9] An organization's global mindset has an important effect on its decisions to go global, its objectives in doing so, and the structures and practices it adopts in order to achieve those objectives. In turn, the results of those efforts influence the dominant mindset. If a company fails in its efforts to create equitable partnerships with its international contacts, but is able to reduce its labor costs, it is likely to adopt a mindset that expands efforts to minimize labor costs, and abandon those to open new markets. Even if the outcome was the result of inappropriate structures or practices, it is easier to attribute it to the characteristics of other cultures, rather than to strategic errors. In this way, the core layer influences the operational and contextual layers of our model, and also is influenced by both of them.

The Operational Layer

An organization's global orientation will be successful only if it is consistent with the operational advantages involved in going global. The operational

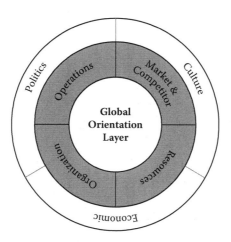

FIGURE 2.3
The operational layer.

layer (Figure 2.3) is composed of three primary factors: global operations and supply chain, global market participation, and global resources.

Global Supply Chain

By definition, the term *operations* refers to activities that directly create organizational value—all the activities from gathering basic raw materials to ultimately delivering a product or service into hands of the end users. In production-oriented firms, one key to operations is the *supply chain*, the flow of goods and information among a network of facilities and transportation system. *Global operations* and supply chain therefore refer to the capability of a company to disperse, geographically, all of its major value-creating activities to various locations around the world in order to take optimal advantage of the resources that are located in those locations. This allows the company to simultaneously deliver lower-cost and higher-quality products and services, through shorter lead time, improved use of resources, and superior response to global customers' requirements.[10] One good example is when General Motors Corporation (GM) allocated its Le Mans operations and supply chain strategies throughout its offshore manufacturing plants and oversea suppliers. GM dispersed many of Le Mans production activities to various countries, where of the $20,000 paid to GM:

- Approximately $6,000 goes to South Korea, where Le Mans is assembled.
- $3,500 goes to Japan for advanced components.
- $1,500 goes to Germany where the Le Mans was designed.
- $800 goes to Taiwan, Singapore, and Japan for other small components.

- $500 goes to Great Britain for advertising and marketing service.
- $100 goes to the Republic of Ireland for data processing.
- The remaining $7,600 goes to GM and to lawyers, bankers, and insurance companies that GM uses in the United States.

As demonstrated above, GM can tap the expertise of many field specialists. Like Mercedes' and BMW's operations in the southern United States, GM can benefit from the assembly work provided by skillful, educated, and industrious Korean workers and at inexpensive cost compared to home. Advanced components are brought from Japan, where GM benefits from local high-end technology, while small parts, which require less complex technology and reduced skills, are outsourced to Taiwan and Singapore, where a great number of low- to mid-technology manufacturers are located. All of the components and small parts come from plants in the western Pacific, near the assembly plant in South Korea. This strategic supply chain helps reduce GM's purchasing complexity and transportation costs. In addition, the design from Germany maintains the Le Mans classic look. Indirect activities such as insurance and banking are conducted in the United States, where the company's headquarters and other supporting functions are located. Even though the Le Mans is a product from an American company, the United States provides very few tangible resources in direct value-creating activities. The globalization of operations and the supply chain helps the company to take advantage of the finest operation practices and the most favorable product costs.[11]

Of course, the Le Mans system creates very complex challenges related to configuration and coordination. In turn, both of these depend on effective communication. *Configuration* refers to the location of the different facilities and their timely access to the resources they need to do their jobs. *Coordination* refers to the processes through which each element in the chain is linked to each other element in a systematic way. A supply chain cannot be configured properly unless information about customer requirements, supplier capacities, and firm capabilities are readily available to organizational decision makers. Similarly, coordination depends on the rapid transmission of accurate information about each element's operations to other elements in the supply chain. Communication breakdowns, or even communication delays, in any link will reduce the effectiveness and efficiency of the system.[12] In global organizations the demands on communication, configuration, and coordination are substantially greater than in domestic-only firms. The different functions are located far away from one another, in terms of both distance and time. Some functions may be outsourced to other multinational firms, removing them from the direct control of the home organization. Modern supply chain systems, such as just-in-time inventory management, place even greater stresses on the supply chain. When configuration, coordination, and communication all work well, the firm will be able to sustain a competitive advantage in fulfilling their customers' expectations. But, if the system breaks down at any point, even for a brief period of time, those expectations will not be met.[13]

Global Market Participation

Global market participation involves a company moving away from treating markets as distinct national entities isolated from one another by trade, distance, time, and culture barriers to treating them as regional or global connections. Globalized market participation allows a company to exploit opportunities in new emerging markets and, at the same time, diversify the risk and uncertainty that accompanies them. For example, in 2001 China consumed about 120 million cellular phones, a market penetration of about 10% of its total population. By June 2007, the country had 501.6 million mobile phone subscribers, according to the Chinese Ministry of Information Industry. The telecommunication firms that move themselves into the Chinese market first should be able to develop a better understanding of local behaviors and consumption patterns, giving them a step up in capturing a major share of this market. Firms that enter the market later will have to overcome those advantages, which will require a substantially greater investment of capital and effort.[14]

However, getting there first is a successful strategy only if firms also "get there smart," that is, only if they carefully select countries on the basis of their potential contribution to the benefits of a globalized market and the impact that those new markets will have on the global competitive position of the business. In addition, market strategies must be adapted to local situations, meaning that a firm's strategy in one country may be very different than its strategy in another. For example, Stanley Works has developed a hammer without claws for carpenters in Central Europe who prefer to use pliers to pull out bent nails, and levels shaped like elongated trapezoids, which the French market prefers. Unilever, a large consumer products company, diversifies its product delivering service worldwide in order to adapt the distribution of its brand to suit local realities. In Europe, it more easily dispatches through online ordering of frozen food. While in Tanzania, it has piloted the bicycle delivery of products to villages inaccessible to motor transport.[15] Similarly, the potential advantages of globalized market risk management also depend on careful strategic decision making. The optimal strategy seems to be balancing risks across global markets.[16]

Balancing global market participation effectively is not easy and requires substantial efforts in worldwide coordination among sales, marketing, and postsales service departments. There are three ways in which firms can coordinate marketing globally. First, a firm may use a single global method across countries to reinforce the company's reputation and brand image. Doing so requires an established product identity or a global communication system for developing that image. Second, firms can transfer learning and experience from one subsidiary to others operating in similar cultures. Making this system work requires some means of allowing subsidiaries to maintain their own autonomy while sharing key information with others. The final approach is to combine these two methods into a centralized information collection and distribution system. The optimal form of market coordination will depend on the particular industry, company, and country because of their particular market conditions.

Global Resources

When organizational decision makers first consider going global, their attention often focuses on obtaining access to low-cost, high-quality resources. These may be tangible, such as raw materials, or labor. We have discussed both of these resources earlier in this chapter and will return to them in the case studies that make up Chapters 2 through 7. However, an additional and important aspect of global resourcing is research and development (R&D).

Traditionally organizations have kept R&D activities close to home, both so that they will be close to organizational decision makers and because of fear that information about new technologies may be "leaked" to competitors. However, today R&D activities are becoming more dispersed. Between 1970 and 1990, 65% of the new laboratories of Fortune 500 firms were located abroad. Overall, between 1987 and 1997 U.S. companies increased their R&D spending abroad from US$5.2 billion to US$14.7 billion. For example, in 1997 Canon carried out R&D activities in five different countries, Motorola did so in seven countries, and Bristol-Myers Squibb in six countries.[17]

There are a number of reasons for this change. Most important, companies need to employ the host country's best engineers and scientists if they are to stay abreast of technological developments in those countries. Overseas R&D facilities also allow companies to better respond to specific requirements of local markets. This is especially important when the cultural and economic gap between the countries is vast. This need pressures the company to seek new knowledge and technology across the globe and bring them back into organization. In addition, offshore R&D operations can significantly reduce the time required to move from research to development to placing products in a market. An obvious example is the semiconductor industry, where the product life cycle is typically less than 6 months. Finally, there may be political pressures from host governments to locate joint-technical activities in host countries. Most governments realize that they will benefit from globalization over the long term only if they can use it to enhance their capacities for value-added activities.[18]

The Context Layer

The final layer of our global model involves the broad context within which modern organizations must operate (see Figure 2.4).

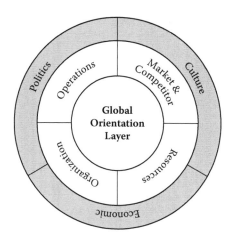

FIGURE 2.4
The context layer.

We will focus our attention on culture for two reasons. First, a society's economic and political system tends to be the tangible expression of its culture. For example, the concept of one person, one vote, which is basic to the U.S. political system, and the system of largely unregulated free markets, which underlies the U.S. economic system, both make sense in a culture that is highly individualistic. However, if those political and economic systems are implemented in cultures with a radically different history, institutions, and underlying values, as it was in Russia after the fall of the Soviet Union, it works far less effectively.[19] Second, when organizations encounter difficulties in their overseas operations, they most often involve problems of intercultural understanding and communication. If intercultural interactions go smoothly, it is relatively easy to deal with political and economic factors. If they go badly, knowledge of a host country's political and economic system is not likely to compensate.

Human beings all experience the same challenges in life, the most important of which involve forming and maintaining mutually satisfying relationships with other people, coping with the passage of time and the effects of aging, and managing pressures imposed by the world around them. Fortunately, we do not have to confront these challenges alone. The people who live around us face the same challenges, and by communicating with them we learn how to make sense out of them and how to deal with them. Different groups develop their own interpretive frameworks and their own patterns of action. These shared systems of interpreting and acting are called *cultures*. Unfortunately, there are almost as many definitions of *culture* as there are people who study them. Sonja Sackman offers one of the most complete definitions:

> A set of assumptions shared by a group of people. The set is distinctive to the group. The assumptions serve as guides to acceptable perception,

thought, feeling, and behavior, and they are manifested in the group's values, norms, and artefacts. The assumptions are tacit among members and are learned by and passed on to each new member of the group.[20]

Geert Hofstede defines *culture* as the "collective programming of the mind which distinguishes the members of one group or category of people from another," and Fons Trompenaars and Charles Hampden-Turner define it as a shared "system of meaning that dictates what we pay attention to, how we act, and what we value."[20] Although somewhat different, each of these definitions has three characteristics in common: (1) culture is composed of a set of assumptions that people accept and use, usually without thinking about them; (2) those assumptions are arbitrary in the sense that any number of different sets of assumptions can be constructed, all of which work quite well in helping the people who accept them manage life's challenges; and (3) they are learned, not innate.

We would add a fourth characteristic, that cultural assumptions are amazingly stable. The members of a society are constantly exposed to messages that support the underlying assumptions of their culture. They learn to interpret ambiguous information in ways that confirm our taken-for-granted assumptions, and tend to ignore or rationalize information that disconfirms them. By learning and accepting the assumptions of a society we become qualified to participate in that society. By learning how people are supposed to think and act, we also accept limitations on how we think and act.[21] We come to view the attributes of our home cultures as *natural*, which implies that they are inevitable, absolute truths, and *normal*, which means that they are morally right, among other things. As a result, they create powerful guidelines and constraints on our actions and the actions of our organizations. As important, they make it very easy for us to take on *ethnocentric* perspectives, and very difficult for us to adopt the *polycentric* perspectives that help us adapt to global organizations. On the other hand, the taken-for-granted assumptions of other cultures guide and constrain how we should act with their members. This is especially important for multinational organizations:

> National culture is a central organizing principle of employees' understanding of work, their approach to it, and the way in which they expect to be treated.... When management practices are inconsistent with these deeply held values, employees are likely to feel dissatisfied, distracted, uncomfortable, and uncommitted. As a result, they may be less able or willing to perform well. Management practices that reinforce national cultural values are more likely to yield predictable behavior, self-efficacy, and high performance.... For managers of multinational corporations, the implications are that adaptation to local cultural conditions is necessary to achieve high performance outcomes.[22]

Being able to adapt to national cultures begins with an understanding of their key characteristics. Although sociologists and anthropologists have developed a number of different models of national cultures, the most influential was that of Geert Hofstede, developed during the 1980s and 1990s. His research with IBM employees in more than fifty countries concluded that national cultures seem to be defined by four dimensions:

1. *Uncertainty avoidance*: The most important dimension, uncertainty avoidance (UA) refers to the extent to which people feel threatened by new, uncertain, or ambiguous situations as well as the ways that people of a particular culture learn to deal with that uncertainty.

2. *Power distance*: The second most important dimension, power distance refers to the extent to which people expect power, wealth, and privilege to be distributed unequally among members of a society or organization, accept those inequalities, and respond to them.

3. *Individualism-collectivism*: In highly individualistic cultures, ties between people are loose. Each person is expected to look out for himself or herself and his or her immediate family. Individual achievement is valued highly, and a person's identity depends on his or her individual skills, abilities, and success. In collectivist cultures, people are defined by their relationships to other people, primarily extended family and close friends or coworkers, and are primarily responsible to those identity groups.

4. *Masculinity-femininity*: This has two dimensions—the extent to which gender roles are clearly defined as separate and distinct from one another, and the particular behavioral expectations associated with gender.

The way these dimensions manifest in the workplace are summarized in Table 2.2.

Early critics of Hofstede's model argued that it did not apply especially well to Asian cultures.[23] Eventually he added a fifth dimension, labeled *Confucian dynamism*, which was a combination of high power distance, taking a long-term orientation toward time, believing that the family is the prototype of all social organizations (and that in good families people suppress their tendencies toward individualism and work to maintain harmony and help everyone "save face"), a reverse golden rule (the belief that one should not treat others as one would not want to be treated), and an emphasis on moderation, frugality, hard work, patience, perseverance, and control of emotions. Hofstede eventually concluded that the most important of these dimensions, and the one that was truly independent of the other four, was *time orientation*, the extent to which people learn to make decisions in terms of long-term concerns, with related values of patience, thrift, perseverance, and respect for tradition, or in terms of short-term considerations, such as efficiency, immediate profits, and a separation of business and family. Of course, each of these dimensions is more complicated than it seems upon

TABLE 2.2

Dimensions Distinguishing Cultures in the Workplace

Small Power Distance Index	Large Power Distance Index
• Hierarchy means role inequalities based on convenience	• Hierarchy means existential inequality
• Subordinates expect to be consulted	• Subordinates expect to be told what to do
• Ideal boss is resourceful democrat	• Ideal boss is benevolent autocrat
Weak Uncertainty Avoidance	**Strong Uncertainty Avoidance**
• Dislike rules	• Emotional need for rules
• Less formalization and standardization	• More formalization and standardization
Collectivist Society	**Individualist Society**
• Value standards differ for in-group and out-group	• Value standards apply to all
• Others are seen as members of group	• Others are seen as resources
• Relationship prevails over task	• Task prevails over relationship
Feminine Society	**Masculine Society**
• Stress on equality, solidarity, and quality of work life	• Stress on equity, competition, and performance
• Managers expected to use intuition, deal with feelings, seek consensus—hold modest career aspirations	• Managers expected to be decisive, firm, assertive, aggressive, just—hold ambitious career aspirations
• Job applicants undersell themselves	• Job applicants oversell themselves
Short-Term Orientation	**Long-Term Orientation**
• Quick results expected; stands for the virtues related to the past and the present, in particular, respect for tradition	• Stands for future rewards, persistence, perseverance
• Protection for one's face	• Face considerations common but considered a weakness

Source: Based on Hofstede (2001), see note 5.

first glance, and they are all interrelated in complex ways. We will examine each dimension in more detail in later chapters, focusing on the dimensions that are most relevant to each case study as it is introduced.

Complicating the Concept of National Cultures

In the social sciences a concept can be both exceptionally influential and highly controversial. This has been true of the idea of national cultures ever since it was first introduced. One of the first concerns was that each of the models that was developed was based on a small sample, and it is very risky to generalize from even a couple of thousand people to an entire nation. This was especially true of Hofstede's research because all of the people in his initial studies also worked for a single company, IBM. Over time, this became less of a problem, for two reasons. Most important, during the 25 years since Hofstede's *Culture's Consequences* was first published, researchers around the world conducted similar research and found similar results. Much of the second edition of his book, and its forty-five-page-long bibliography, is devoted to summarizing this research. His model also has been changed in a number of ways to respond to critics. Introducing the concept of Confucian dynamism and time orientation was the most important revision. The result has been a more complicated model, one that better reflects the ways in which culture and cultural differences complicate global engineering.

Some Nations Are Mazes of Subcultures

One complication involves the concept of *national* cultures. Some nations are much more culturally homogeneous than others. For example, Iceland and Japan have long and rich histories, and their populations are made of people with very similar backgrounds, experiences, beliefs, and values. Other nations are like China—a combination of very different cultures that have been unified for thousands of years, long enough to develop some shared experiences, beliefs, and values, but still composed of a number of distinct subcultures. Still other "nations" are composed of very different peoples, and are nations only because outside forces or powerful internal leaders have dictated that they unified. For example, Yugoslavia was a unified nation of Croats, Serbs, Albanians, Macedonians, and Bosnians for decades largely because of Marshall Tito's power. Iraq is a separate nation largely because the British declared that it was, and was able to enforce that declaration militarily, not because Sunnis, Shiites, and Kurds are culturally homogeneous peoples. Similarly, Kurdistan is not a separate nation because outside powers decided to draw borders in a way that placed some Kurds in Iraq, others in Turkey, and others in Iran. So, while it still is useful to think of cultures as being national in scope, it is even more important to recognize that cultures also differ from one another in terms of how homogeneous they are. Consequently, a multinational company would find it much easier to prepare its employees to understand, adapt to, and work in Icelandic culture than to prepare them to understand, adapt to, and work in postwar Iraq.

Globalization Changes Cultures

A second complication for the concept of national culture has resulted from globalization itself. Globalization rarely has the same effects on every part of a nation. Areas of Mexico that border on the United States have changed much more than areas in the south. The Chinese government made a conscious strategic decision to focus on international trade and the development of private sector organizations in its southern provinces. As a result, globalization has had a much greater impact there than it has in northeastern China. It also has had very different effects on rural China than it has on urban areas like Shanghai, and on workers currently in their twenties and thirties than on other age groups.[24] The high-tech centers of Hyderabad and Bangalore now are very different cultures than any other parts of India. In Turkey, globalization has created two distinct cultures. Managers in the public sector hold on to traditional values that go back to the time of the Ottoman empire. As Hofstede found in his studies of Turkish managers, they are high in power distance, high in uncertainty avoidance, and highly collectivist rather than individualists. They resist change, prefer authoritarian and centralized organizational structures, follow rules, and are reticent about taking initiative. In contrast, Turkey's private sector has grown massively since 1980. Turkish markets have been opened to foreign firms, state-owned industries have been closed or privatized, and competition has increased. As a result, managers in Turkey's private sector are dynamic, value risk taking, are entrepreneurial and innovative, ambitious, flexible, and oppose excessive rules and centralization of decision making and control.[25] Many national cultures have become hybrids so that some areas are more like the globalized sectors of other countries than they are like nonglobalized areas of their own.

Organizations Have Cultures of Their Own

Finally, as some early critics of Hofstede's reliance on IBM employees pointed out, the distinctive cultures of multinational corporations interact with the values of national cultures. Cultures are defined by core *values*, *beliefs*, and *taken-for-granted ways* of viewing reality. When people enter organizations, they bring those cultural assumptions with them. This is the primary reason why organizations within a particular society are very much alike. If the institutions of a society teach employees that bureaucracy is the normal and natural way for an organization to function, or that leaders are supposed to think and followers are supposed to follow orders without thinking, they will expect their organizations to have those characteristics. However, organizations also engage in a great deal of effort to teach new employees that they expect them to think and act in certain ways. These organizational *practices* tend to be consistent with the core values of the societies surrounding the organization; they also define the organization as a unique entity. Even though IBM, Dell, Hewlett-Packard/Compaq, Apple, and Microsoft all are U.S. companies and even operate in the same industry, they have very different organizational cultures, very different practices.[26]

When an organization operates in multiple nations, its employees come from many different cultural backgrounds and bring many different sets of values, beliefs, and ways of viewing reality with them to the organization. These differences create tensions and incongruities that the organization must manage in some ways. Of course, the simplest solution is to dictate that every division of Company X will operate in precisely the same way, regardless of the culture surrounding each branch. The company places the burden of adapting on employees from or in societies other than the one surrounding the main office. Hofstede and others have noted that this is the approach typically used by U.S. multinationals, regardless of where a particular operation is located.[27] The practices of the organization, and indirectly, of the home society, are given priority over the cultural values of the various nations in which it operates.

For example, one of the most common practices of U.S. organizations is some form of pay for performance. The simplest version is the piece-rate system that became popular in U.S. manufacturing firms around 1900—each employee is paid a certain amount for each unit that he or she produces. Most pay-for-performance systems are much more complex than this example suggests, especially for workers who are not involved in manufacturing. In addition, there is little evidence that U.S. firms actually use pay-for-performance systems with their executives.[28] But, such systems are easy to defend at home because they are very consistent with the individualism of U.S. culture— each employee's rewards are based on his or her *individual* performance— and applying the system *universally*—to all employees regardless of their individual situations—corresponds to U.S. definitions of *fairness*.[29] However, in highly community-oriented cultures, universal systems are inherently unethical and unfair because they do not take relational considerations into account. Trompenaars and Hampden-Turner describe a U.S. computer company, which they give the pseudonym MCC, that persisted in universal application of a pay-for-performance system in spite of its failure. During the first year of a two-year pilot study with the company's sales force, workers spent a great deal of time and energy discussing the system. By the end of the first year, one-third of the employees left the company because they felt that the system treated them unfairly. Total sales remained flat. Trompenaars and Hampden-Turner note that "despite this disaster, management continued this experiment because they believed that this kind of change was necessary and would take time to be accepted."[30] When the system was applied in the Middle East, the system worked briefly, because workers valued seeing their coworkers succeed, as one would expect in community-oriented cultures. But, once they learned that the vast majority of workers had lost income while a small number received significant rewards, morale and sales plummeted. When supervisors learned that some of their subordinates were making more than they were, they were incensed. In their high power distance culture, formal rank and status were supposed to mean higher income. Worse yet, the incentive system encouraged salesmen to load their clients up with products that they could not sell, undermining those interpersonal relationships. Eventually, the system was abandoned.

Unfortunately, although managerial fads come and go, almost all of them are developed within the United States or in similar cultural contexts. It does not really matter what the fad is—from management by objectives (MBO) to matrix organizations to systems of participatory decision making (PDM)—it is likely to be grounded in individualistic, universalistic cultural assumptions.[31] Ironically, when companies attempt to impose practices derived in their home cultures on employees from other cultures, everyone's attention is focused on cultural *differences*. As a result, the likelihood of cross-cultural conflict is magnified and the ability to develop approaches that are appropriate to both cultures is reduced.

Eventually, employees do find ways to cope with the incongruities. Trompenaars and Hampden-Turner describe the experience of Geddy Teok, who after 4 years of careful negotiating had finally won acceptance of a joint venture arrangement with a Japanese firm. One week before the signing ceremony was to be held, Geddy received an eleven-hundred-page contract from the home office, located in a very universalist, rule-oriented culture. He knew that the partners would view the contract as evidence that his company did not really trust their Japanese counterparts, and would refuse to sign. He called the legal department in the home office, hoping to get them to adapt to the cultural realities that he faced. But, they told him that the contract was even longer than usual because they had been told that Asians in general and Japanese in particular had a reputation of being quite loose in defining what was developed by them and what came from outside. So, the legal department wanted to cover every possible contingency in writing. Geddy knew that his job was on the line—if he presented the complete contract to the Japanese firm he would fail to get the contract; if he reduced it to a couple of pages and presented it as a symbol of their mutual trust, he would have to deal with his legal department. He decided to tell his Japanese counterparts the truth—that extended contracts are normal business practices in the United States, and are not meant to be insulting, and that he was confident that they could work together to resolve any legal issues that might arise. The Japanese response was, "How long will you stay here, Mr. Teok?" Geddy responded that he would stay until the job was done, and the contract was signed.[32]

In the MCC example, expatriate managers learned to work backwards, first determining their employees' rewards based on the traditional practices that were accepted in their cultures. Then, they played games with the pay-for-performance system so that they could justify the decisions that they already had made in terms of a measure of performance that was acceptable to the home office. Similarly, Muslim employees operating in U.S. firms learn to live "split" lives, adopting U.S. practices and behaviors at work while maintaining traditional lives outside of work.[33] In all three cases the culturally normal practices of the headquarters of multinational firms—requiring members of other cultures to adapt to standard procedures instead of adapting those procedures to other cultures—were dominant, but also were overcome.

Other firms place the burden of adapting on its home office employees, and on expatriates that are assigned to posts in other nations. This means that these assignments can be very stressful, and require a lengthy experience in the host society. The problems that employees encounter are different, but also can be managed. Even organizations who initially fail in cross-cultural operations often learn from their mistakes, begin again, and eventually succeed.[34]

The Political and Economic Dimensions

A society's culture, political system, and economic model are interrelated in complex ways. Societies create political systems that they believe are appropriate to the core values and beliefs held by their members, and that assist the development of the economic value that is most likely to support the kind of society that citizens want to be part of. In turn, the economic model a society creates makes some values more visible and more important than others, and may even lead a society to change its core values. Similarly, political systems tend to represent the interests of some groups of people more effectively than other groups, and those inequities may lead citizens to reevaluate their core values or to change their economic systems. Organizations function most efficiently if they are adapted to the political and economic realities within which they operate. But, organizations also wield significant political and economic power, and at least over the long term can influence cultural values.

Perhaps an example will explain these ideas. In the United States, organizations tend to be large, privately owned, and, when compared to other countries, only loosely regulated by government. This was not always so. Two hundred years ago there were very few large organizations. In the 1890s there were only a few, and they operated railroads or manufactured textiles or steel. Suddenly, things changed—within 5 years the two hundred biggest corporations of the time were formed, many of which still exist. There are a number of reasons for this shift—the United States was rich in natural resources, it had access to a seemingly inexhaustible supply of inexpensive labor from Europe, its culture celebrated individual achievement and entrepreneurship, and it had a massive market for goods that suddenly had become easily accessible because of improving transportation and communication technologies. The rapid industrialization and economic growth following the Civil War concentrated wealth in the hands of a much smaller number of people than had ever been true in U.S. history. The economic power of these "robber barons" quickly exceeded that of other institutions. The United States was settled by immigrants, who brought with them a deep distrust of the dominant institutions of Europe and Asia—the official church and government—and that distrust made appeals to limit government very powerful. However, corporations were a very new kind of institution and there was a great deal of mistrust in them as well. As a result, the first laws governing corporations required them to be operated in the interests of the public as a whole and to have public representatives on their boards of directors.

But, in 1819 a group of New England businessmen, represented by Daniel Webster, persuaded the U.S. Supreme Court to rule that those requirements were unconstitutional. In the same year, owners of the largest corporations persuaded courts and legislatures to create limited liability, a concept that says that if a company goes bankrupt and cannot pay its creditors or workers, its managers and owners do not have to pay its debts out of their own pockets, regardless of how wealthy they might be. The Supreme Court also declared that only the federal government had the power to regulate inter-state commerce. This was an important change in the U.S. political system because at the time the federal government was much less powerful than the governments of the largest states. Large corporations found it difficult to dominate powerful state governments, but could easily dominate the federal government. Finally, owners obtained the legal right to sell stock in their companies, and to do so in a way that limited individual stockholders' influence over their decisions or operations. This legal change gave them almost unlimited access to the funds they needed to enlarge their organizations, and with increased size came increased political influence. Taken together, these actions gave the executives of U.S. firms a unique ability to increase the size and independence of their organizations, more than in any other society at the time. Of course, creating the kind of economic system they desired was not easy. Owner-managers had to persuade policy makers that the changes they preferred would provide the kind of economy that was valued in U.S. society—one that was efficient, effective at generating jobs, and generally more capable of meeting the overall society's needs. The fact that their firms often were *not* more efficient, more productive, or more socially responsible than government or religious organizations is testimony to the success of their persuasive appeals. The result was a unique American system, a dis-tinctive combination of political structures, economic systems, and cultural values.[35] Suddenly, small organizations were placed at a competitive disad-vantage. They also had to grow or die. Within a decade, the large corporation became the dominant strategy for U.S. organizations. For the next 30 years the U.S. economy boomed, so the large, unregulated corporation seemed to be the best strategy. Eventually the superiority of the system became a taken-for-granted assumption of U.S. culture. It no longer needed to be justified, because there did not seem to be any viable alternative.

Other societies developed other combinations of political systems, eco-nomic models, and organizational strategies, combinations that are consis-tent with the values and assumptions of their cultures.[36] As long as their organizations operate primarily within their borders, people are not likely to even think about their own systems. Their governments can develop ways of insulating their economies from external pressures, so the local political-economic systems continue to seem to be optimal.

But, in a global economy people encounter organizations from societies that have different political-economic-organizational systems. They sud-denly realize that their ways of doing things are the result of choices that have been made in the past and that they repeat every day. And, they may discover that they have no choice but to adapt to those alien ways of think-ing and acting. In a global economy, even governments may not be able to protect their organizations from external pressures. More than half of the world's one hundred largest economies are corporations, not countries. The two hundred largest corporations account for one-third of the world's eco-nomic activity, although they employ only three-fourths of 1% of its people.

The five hundred largest corporations account for 70% of world trade. No government can ignore that kind of economic power; only a handful can resist it. MIT economist Lester Thurow concludes:

> The knowledge-based economy is fundamentally transforming the role of the nation-state. Instead of being a controller of economic events within its borders, the nation-state is increasingly having to become a platform builder to attract global economic activity to locate within its borders.... Because countries need corporations more than corporations need countries, the relative bargaining power of governments and multinational corporations is shifting. High profile multinational companies ... no longer pay taxes to governments. Governments pay taxes to them.[37]

Just as cities and states within the United States long have competed with one another to attract new industries, nations now compete to attract multinational corporations. Israel paid $600 million (in grants, facility construction, and tax rebates) to land an Intel plant, Brazil promised to pay Ford $700 million, and so on. Similarly, the tax burden felt by corporations has declined significantly throughout the developed world. In the late 1950s approximately one-half of U.S. federal income tax receipts came from corporations; by 1996 that figure had fallen to around 10%, and by 2003 it was 6%. In 2003 the 275 most profitable U.S. organizations did not even have to report one-half of their US$1.1 trillion profits to the Internal Revenue Service. Financier George Soros concludes that worldwide, the proportion of a country's income that goes to taxes has not changed substantially in recent years, but as political power has shifted to large, multinational corporations, "the taxes on capital and employment have come down while other forms of taxation, particularly on consumption, have been increasing. In other words, the burden of taxation has shifted from capital to citizens."[38] Of course, the pace of these changes varies across countries and regions. As a result, there still are significant differences in the political and economic systems in place in different countries. These differences complicate the lives of employees of global organizations in myriad ways.

Conclusion

In this chapter we have discussed a framework that can be used by engineers to ensure a sound global perspective when dealing with international projects. The global engineering model presented will prove useful to engineers that are not typically accustomed or trained to deal with nontechnical decisions. In the next chapter, we introduce a decision methodology that will aid the engineer to make technical decisions involving both quantitative and qualitative criteria.

Review and Study Questions

1. Consider the examples of companies given in the beginning of Chapter 1. Identify the main points of each example, and classify them using GEM.

2. In some countries in Asia you will find that production personnel are less willing to openly discuss and give ideas for the improvement of the system or organization they work in. One of the reasons for this is that they feel that in doing so they would be admitting there are problems with the system or organization, and as a result they would be admitting that their supervisor, whom they respect very much, is not doing an excellent job. This situation constitutes a problem for the implementation of continuous improvement programs where the personnel must routinely identify and suggest ideas for improvement. What layer of GEM would contain the important issues in this case?

3. According to Hofstede's experimental results, Japan has a higher uncertainty avoidance index than Hong Kong (the higher the index, the more uncertainty avoidance). In which country would you expect to implement standardized procedures?

4. According to Hofstede's experimental results, Denmark has a small power distance index and Malaysia has a large power distance index. Consider that there are two good managers that need to be deployed to these countries: Manager A can be described as a resourceful democrat and Manager B is known for being a benevolent autocrat. What manager should be assigned to which country?

5. According to Hofstede's experimental results, Singapore has a small individualism index and the United States has a large individualism index. In which country would you expect to have more team start-up problems when assigning new teams to execute projects?

Notes

1. Charles Conrad and M. Scott Poole, *Strategic Organizational Communication*, 6th ed. (Belmont, CA: Wadsworth, 2005); S. H. Rhinesmith, *A Manager's Guide to Globalization: Six Keys to Success in a Changing World* (New York: Business One-Irwin/ASTD, 1993).
2. B. Chakravarthy and H. V. Perlmutter, "Strategic Planning for a Global Economy," *Columbia Journal of World Business* (Summer, 1985): 3–10.

3. Fons Trompenaars and Charles Hampden-Turner, *Riding the Waves of Culture* (New York: McGraw-Hill, 1998); Y. Zeirra and E. Harari, "Host-Country Organizations and Expatriate Managers in Europe," *California Management Review* 21 (1979): 40–50.

4. V. Govindarajan and K. A. Gupta, *The Quest of Global Dominance: Transforming Global Presence into Global Competitive Advantage* (San Francisco: Jossey-Bass, 2001).

5. Geert Hofstede, *Culture's Consequences*, 2nd ed. (London: Sage, 2001), especially pp. 400–445.

6. P. Evans, V. Pucik, and J. L. Barsoux, *The Global Challenge: Frameworks for International Human Resource Management* (Boston: McGraw-Hill/Irwin, 2002).

7. Govindarajan and Gupta, *Quest of Global Dominance*; B. L. Kedia and A. Mukherji, "Global Managers: Developing a Mindset for Global Competitiveness," *Journal of World Business* 34 (1999): 230–51; Noel M. Tichy, "Global Development," in *Globalizing Management: Creating and Leading the Competitive Organization*, ed. V. Pucik, N. M. Tichy, and C. K. Bartlett (New York: John Wiley & Sons, 1992).

8. Conrad and Poole, *Strategic Organizational Communication*; Cynthia Stohl, "Globalizing Organizational Communication," in *New Handbook of Organizational Communication*, ed. F. Jablin and L. Putnam (Thousand Oaks, CA: Sage, 2000).

9. M. L. Dertouzos, R. S. Lester, and R. M. Solow, *Made in America: Regaining the Productive Edge* (New York: Harper Perennial, 1989).

10. E. Frazelle, *Supply Chain Strategy* (New York: McGraw-Hill, 2002).

11. C. W. Hill, *International Business: Competing in the Global Marketplace* (Burr Ridge, IL: Irwin, 1994), p. 6.

12. See Conrad and Poole, *Strategic Organizational Communication*; M. E. Porter, "Changing Patterns of International Competition," *California Management Review* 28 (1986): 9–40; and S. E. Fawcett, L. Birou, and B. C. Taylor, "Supporting Global Operations through Logistics and Purchasing," *International Journal of Physical Distribution & Logistics Management* 23 (1993): 3–11.

13. Fawcett et al., "Supporting Global Operations"; R. Hayes, S. C. Wheelwright, and K. B. Clark, *Dynamic Manufacturing: Creating the Learning Organization* (New York: Free Press, 1988).

14. Govindarajan and Gupta, *Quest of Global Dominance*; Hill, *International Business*.

15. S. H. Kale and R. P. McIntyre, "Distribution Channel Relationships in Diverse Cultures," *International Marketing Review* 8 (1991): 31–45; Porter, "Changing Patterns of International Competition"; A. M. Rugman and R. M. Hodgetts, *International Business: A Strategy Management Approach* (New York: McGraw-Hill, 2000); www.unilever.com, 2002.

16. B. Lloyd, "The Outlook for Globalization," *Leadership & Organization Development Journal* 17 (1996): 18–23; G. S. Yip, *Total Global Strategy: Managing for Worldwide Competitive Advantage* (Englewood Cliffs, NJ: Prentice-Hall, 1992).

17. Yamashina Kuemmerle, "Challenge to World-Class Manufacturing," *International Journal of Quality & Reliability Management* 17 (1997): 132–43; V. Terpstra, "International Product Policy: The Role of Foreign R&D," *Columbia Journal of World Business* 12 (1977): 24–32; D. H. Dalton and M. G. Serapio, Jr., *Globalizing Industrial Research and Development* (U.S. Department of Commerce, Office of Technology Policy, September 1999), pp. 7–9, 53–54; V. Chiesa,

"Evolutionary Patterns in International Research and Development," *Integrated Manufacturing Systems* 7 (1996): 5–15; R. D. Pearce and S. Singh, *Globalizing Research and Development* (London: Macmillan, 1992).

18. Govindarajan and Gupta, *Quest of Global Dominance*; O. Granstand, A. Hakanson, and S. Sjolander, eds., *Technology Management and International Business* (Chichester, England: John Wiley & Sons, 1992); Kuemmerle, "Challenge to World-Class Manufacturing."

19. George Soros, *The Crisis of Global Capitalism* (New York: Public Affairs, 1998), especially chapter 3.

20. Sonja Sackman, ed., *Cultural Complexity in Organizations: Inherent Contrasts and Contradictions* (Thousand Oaks, CA: Sage, 1997). Also see Kevin Phillips, *The Politics of Rich and Poor* (New York: Broadway Books, 1990) and *Wealth and Democracy* (New York: Broadway Books, 2002); Hofstede, *Culture's Consequences*; and Trompenaars and Hampden-Turner, *Riding the Waves of Culture*, p. 13.

21. Goran Therborn, *The Ideology of Power and the Power of Ideology* (London: Verso, 1980); and Dennis Mumby, *Power in Organizations* (Norwood, NJ: Ablex, 1988), especially chapter 4.

22. Karen L. Newman and Stanley D. Nollen, "Culture and Congruence," *Journal of International Business Studies* 27 (1996): 755.

23. Both criticisms and follow-up research are summarized in Hofstede, *Culture's Consequences*, chapter 7.

24. John Child, *Management in China during the Age of Reform* (Cambridge, UK: Cambridge University Press, 1994); C. C. Ching, "The One-Child Family in China," *Studies in Family Planning* 13 (1982): 208–12; J. Hassard, J. Sheehan, and J. Morris, "Enterprise Reform in Post-Deng China," *International Studies of Management & Organization* 29 (1999): 107–36.

25. Yasemin Arbak, Ceyhan Aldemir, Omur Timurcanday Ozmen, Alev Ergenc Katrinli, Gulem Atabay Ishakoglu, and Julide Kesken, "Perceptual Study of Turkish Managers' and Organizations' Characteristics: Contrasts and Contradictions," in *Cultural Complexity in Organizations: Inherent Contrasts and Contradictions*, ed. Sonja Sackman (Thousand Oaks, CA: Sage, 1997), pp. 87–104.

26. For an extended discussion of these ideas, see Conrad and Poole, *Strategic Organizational Communication*.

27. Hofstede, *Culture's Consequences*, p. 441.

28. Rakesh Khurana, *Searching for a Corporate Savior* (Princeton, NJ: Princeton University Press, 2002).

29. Trompenaars and Hampden-Turner, *Riding the Waves of Culture*. Perceptive readers will notice that there is an important inconsistency between the values of individualism and universalism, which should not be surprising because all cultures have incongruent values and characteristic ways of managing those inconsistencies. For an analysis of how U.S. firms manage this inconsistency, see pp. 31–40. There also are a number of difficulties in implementing pay-for-performance systems, even within highly individualistic cultures. For a summary, see Conrad and Poole, *Strategic Organizational Communication*, especially chapter 3.

30. Trompenaars and Hampden-Turner, *Riding the Waves of Culture*, p. 31. Persistence in the face of evidence of failure also is characteristic of managerial decision making in U.S. firms. See Alan Tegar, *Too Much Invested to Quit* (New York: Pergamon, 1980), for example.

31. For an analysis of managerial fads, see Timothy Clark and David Greatbatch, "Management Fashion as Image-Spectacle," *Management Communication Quarterly* 17 (2004): 396–424. For a cultural critique of recent managerial fads, see Hofstede, *Culture's Consequences*, especially chapters 8 and 9, and Trompenaars and Hampden-Turner, *Riding the Waves of Culture*, especially chapter 12.

32. Trompenaars and Hampden-Turner, *Riding the Waves of Culture*, pp. 45–49.

33. K. M. Alkhazraji et al., "The Acculturation of Immigrants to U.S. Organizations," *Management Communication Quarterly* 11 (1997): 217–265.

34. See Trompenaars and Hampden-Turner, *Riding the Waves of Culture*. Chapters 2 and 3 of this book will discuss this learning process in more detail, because both of the case studies they discuss involve initial failures followed by change and eventual success.

35. Neil Fligstein, *The Transformation of Corporate Control* (Cambridge, MA: Harvard University Press, 1990) and *The Architecture of Markets* (Princeton, NJ: Princeton University Press, 2001); Charles Perrow, *Organizing America* (Princeton, NJ: Princeton University Press). Perhaps the clearest example of these processes is the development of very different railroad systems in the United States, UK, and France. For an analysis, see Perrow and William Roy, *Socializing Capital: The Rise of the Large Industrial Corporation in America* (Princeton, NJ: Princeton University Press, 1997).

36. For example, see Graham K. Wilson, *Business and Politics: A Comparative Introduction*, 2nd ed. (Chatham, NJ: Chatham House Publishers, 2001).

37. Lester Thurow, "Globalization," *Annals of the American Academy of Political and Social Science* 570 (2000): 21, 22. Also see L. Thurow, "In Praise of Cultural Imperialism?" *Foreign Policy* 107 (Summer 1997) and "New Rules," *Harvard International Review* (Winter 1997/98): 7–42; and Robert D. Kaplan, "Was Democracy Just a Moment?" *The Atlantic Monthly* (December 1997).

38. Lynnley Browning, "Study Finds Accelerating Decline in Corporate Taxes," *New York Times on the Web* (September 23, 2004); George Soros, *The Crisis of Global Capitalism* (New York: Public Affairs, 1998), p. 112. Also see G. Soros, *The Open Society* (New York: Public Affairs, 2000).

3

Decision Making in Global Projects

Introduction

A characteristic of decisions that involve global projects is that they must take into account multiple dimensions. As described in Chapter 2, these dimensions can be quantitative, qualitative, or both, presenting a special challenge to engineers that are trained to deal with mostly quantitative models. This chapter describes a decision-making methodology that will allow engineers to consider simultaneously multiple qualitative and quantitative factors. The proposed methodology will integrate the analytical hierarchical process (AHP) with the global engineering model (GEM) described in Chapter 2.

Analytical Hierarchical Process

In deterministic decision making there is no uncertainty associated with the data used to evaluate the different alternatives. The analytical hierarchical process (AHP)[1] is a deterministic decision-making approach that allows the incorporation of subjective judgment into the decision process. The decision maker will quantify his or her subjective preferences, feelings, and biases into numerical *comparison weights* that are used to rank the decision alternatives. Another advantage of AHP is that the *consistency* of the decision maker's judgment is also quantified as part of the analysis.

The basic AHP model consists of the *alternatives* that are to be ranked, *comparison criteria* that will be used to rank the alternatives, and a *decision*. Figure 3.1 shows a single-level decision hierarchy with c criteria and m alternatives. By inserting additional levels of criteria, the same model can be applied recursively to form multiple-level hierarchies. For clarity, the following discussion applies to a single-level hierarchy model.

The *objective* of the procedure is to obtain *rankings*, R_j, for each alternative $j = 1, ..., m$, that represent a ranking of the alternatives based on the importance that the decision maker has attributed to each criteria.

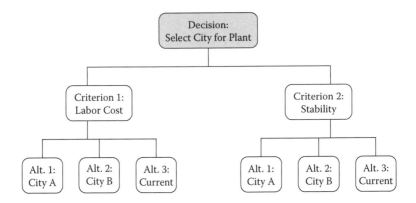

FIGURE 3.1
One-level decision hierarchy.

The *comparison matrix, A* = [a_{rs}], is a square matrix that contains the decision maker's preferences between pairs of criteria (or alternatives). AHP uses a discrete scale from 1 to 9, where a_{rs} = 1 represents no preference between criteria, a_{rs} = 5 means that the row criterion *r* is strongly more important than the column criterion *s*, and a_{rs} = 9 means that criterion *r* is extremely more important than criterion *s*. For consistency, a_{rr} = 1 (i.e., comparison against itself), and a_{rs} = 1/a_{sr}.

The *normalized comparison matrix, N* = [n_{rs}], normalizes the preferences in matrix *A* such that the column sums add to 1.0. This is obtained by dividing each entry in *A* by its corresponding column sum. If *A* is a $q \times q$ matrix, then the elements of *N* are

$$n_{rs} = \frac{a_{rs}}{\sum\limits_{k=1}^{q} a_{ks}}$$

The *weight, w_r*, associated with criterion *r* is the row average calculated from matrix *N*, or

$$w_r = \frac{\sum\limits_{s=1}^{q} n_{rs}}{q}$$

The AHP calculations to determine the rankings R_j are:

Step 1. Form a $c \times c$ comparison matrix, *A*, for the criteria.

Step 2. Form $m \times m$ comparison matrices for the alternatives with respect to each criterion, or A_i, for i = 1, ..., c.

Step 3. Normalize the comparison matrices obtained in steps 1 and 2. Denote these normalized matrices N and N_i, $i = 1, ..., c$.

Step 4. Determine the weights for criteria and alternatives. Denote these weights w_i and w_{ij} for $i = 1, ..., c$ and $j = 1, ..., m$.

Step 5. Determine the rankings for each alternative

$$R_j = \sum_{i=1}^{c} w_i w_{ij}, j = 1, ..., m.$$

Step 6. Select the alternative with the highest ranking.

The *consistency* of the comparison matrix A is a measure of how coherent was the decision maker in specifying the pair-wise comparisons. For a $q \times q$ comparison matrix A, the *consistency ratio* CR is calculated as follows:

$$CR = \frac{q(q_{max} - q)}{1.98(q-1)(q-2)}$$

where

$$q_{max} = \sum_{s=1}^{q} \left(\sum_{r=1}^{q} a_{sr} w_r \right)$$

Comparison matrices with values of $CR < 0.1$ have acceptable consistency, 2×2 matrices are always perfectly consistent, and matrices with $q_{max} = q$ are also perfectly consistent.

Application Example: Decision Making under Certainty Using AHP

This application example is adapted from Leon.[2] Consider the problem of opening a facility in a foreign country. The alternatives are open in Country A, open in Country B, or keep the current facility, C (see Figure 3.1). For this example the criteria for decision are labor cost and region stability.

First, the decision maker expresses her preferences among criteria in the comparison matrix A, and its corresponding normalized matrix (see Table 3.1).

TABLE 3.1

Comparison and Normalized Matrices between Labor and Stability

a_{rs}	Labor	Stability	n_{rs}	Labor	Stability
Labor	1	4	Labor	0.8	0.8
Stability	¼	1	Stability	0.2	0.2

Next, the decision maker must generate comparison matrices among each alternative with respect to each criterion illustrated in Table 3.2. With respect to labor, the comparison matrix A_{labor} and normalized matrix are:

TABLE 3.2

Comparison and Normalized Matrices between Countries with Respect to Labor

a_{rs}	Country A	Country B	Current Facility	n_{rs}	Country A	Country B	Current Facility
Country A	1	2	7	Country A	0.61	0.63	0.54
Country B	½	1	5	Country B	0.30	0.31	0.38
Current facility	1/7	1/5	1	Current facility	0.09	0.06	0.08

With respect to stability, the comparison matrix $A_{stability}$ (Table 3.3) and normalized matrix are:

TABLE 3.3

Comparison and Normalized Matrices between Countries with Respect to Stability

a_{rs}	Country A	Country B	Current Facility	n_{rs}	Country A	Country B	Current Facility
Country A	1	1/5	1/8	Country A	0.07	0.03	0.10
Country B	5	1	1/6	Country B	0.36	0.14	0.13
Current facility	8	6	1	Current facility	0.57	0.83	0.77

Next, the weights associated with each criterion and alternative are the row averages of the normalized matrices: $w_{labor} = 0.8$, $w_{stability} = 0.2$, $w_{labor,A} = 0.59$, $w_{labor,B} = 0.33$, $w_{labor,C} = 0.08$, $w_{stability,A} = 0.07$, $w_{stability,B} = 0.21$, and $w_{stability,C} = 0.72$. The weights are illustrated in Figure 3.2.

The rankings for each alternative are calculated as follows:

$$R_{country\ A} = w_{labor}\ w_{labor,A} + w_{stability}\ w_{stability,A} = (0.8)(0.59) + (0.2)(0.07) = 0.49$$

$$R_{country\ B} = w_{labor}\ w_{labor,B} + w_{stability}\ w_{stability,B} = (0.8)(0.33) + (0.2)(0.21) = 0.31$$

$$R_{current} = w_{labor}\ w_{labor,C} + w_{stability}\ w_{stability,C} = (0.8)(0.08) + (0.2)(0.72) = 0.21$$

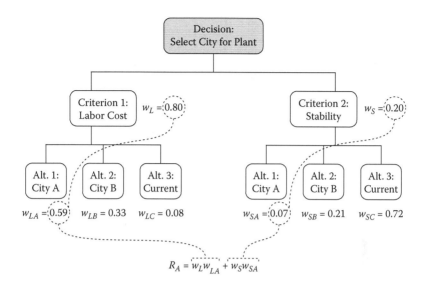

FIGURE 3.2
Rank weights for the example.

AHP suggests building a plant in Country A because it is the alternative with the largest ranking.

The decision maker may desire to quantify how consistent each comparison matrix is. The 2 × 2 criteria comparison matrix is perfectly consistent.

A nice feature of AHP is that it allows the decision maker to evaluate how consistent his or her judgment was when completing the pair-wise comparison matrices. This is accomplished by calculating the consistency ratio CR. The consistency ratio CR is calculated for the 3 × 3 matrices as follows:

Labor: $q_{max} = 0.59(1 + 1/2 + 1/7) + 0.33(2 + 1 + 1/5) + 0.08(7 + 5 + 1) = 3.07$; the consistency ratio:

$$CR_{labor} = \frac{3(3.07 - 3)}{1.98(3 - 1)(3 - 2)} = 0.05$$

Stability: $q_{max} = 0.07(1 + 5 + 8) + 0.21(1/5 + 1 + 6) + 0.72(1/8 + 1/6 + 1) = 3.42$

$$CR_{stability} = \frac{3(3.42 - 3)}{1.98(3 - 1)(3 - 2)} = 0.32$$

Hence, the comparison matrix with respect to labor, A_{labor}, has acceptable consistency (i.e., $CR_{labor} < 0.1$); however, the comparison matrix with respect to stability is inconsistent (i.e., $CR_{stability} > 0.1$), so the decision maker must try

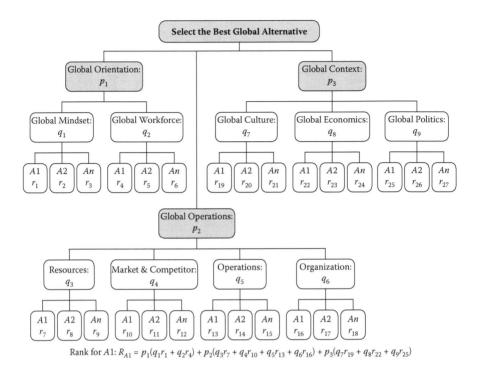

Rank for A1: $R_{A1} = p_1(q_1 r_1 + q_2 r_4) + p_2(q_3 r_7 + q_4 r_{10} + q_5 r_{13} + q_6 r_{16}) + p_3(q_7 r_{19} + q_8 r_{22} + q_9 r_{25})$

FIGURE 3.3
Integrated AHP and global decision model.

to reassess the ratings given in matrix $A_{\text{stability}}$ (see questions for discussion at the end of the chapter).

Integrating GEM and AHP

It is possible to apply AHP in the context of GEM by considering the different layers of GEM as decision criteria. Each layer can be further decomposed into its components by adding additional hierarchy levels to the decision tree, as illustrated in Figure 3.3.

Data Gathering

Perhaps the most challenging aspect of applying the combined AHP-GEM, especially for engineers, would be to assign numerical preference values when comparing qualitative factors that will often appear in the global mindset and context layers of GEM. Fortunately, in addition to expert judgment, it is possible to find useful information in the open literature and the Internet.

Hofstede[3] provides numerical indices for the different cultural dimensions. A summary of this information is given in Table 3.4.

Care should be taken to convert the information contained in Hofstede's graphs into the 1–9 preference scale used in AHP, since a low value of an index does not correspond to a low value in the preference scale. For instance, let's assume you are comparing the UK with Brazil with respect to power distance. The power distance index (PDI) for the UK, PDI(UK) = 35 (relatively low) and PDI(Brazil) = 69 (relatively high). In a situation where having low PD is strongly preferable, you would give the UK a high preference value compared to Brazil. Conversely, in a situation where it is strongly preferable to have high PD, you would give Brazil a high preference value compared to the UK. Hence, the mapping of numerical information (indices or rankings) into the preference scale will depend on the decision at hand.

There is a vast amount of information on the Internet that can be used to compare different countries. For instance, some organizations rank different countries with respect to their logistics capabilities, or report on how different countries are perceived with respect to corruption; see for instance, http://www.transparency.org/publications/publications/ar_2006_special_asia_pacific. An example of such rankings is given in Table 3.5.

Example: MexPlast Goes Global

MexPlast is a Mexican company that is owned by a group of young Mexican entrepreneurs. The company has been in business for 8 years, and has been very successful selling specialty plastic molded parts in the Mexican market. Most people working in the company have been carefully recruited from the best universities in Mexico, and the operators carefully selected from technical schools. Now that MexPlast has a solid national operation, the company is seeking to buy or partner with foreign manufacturing companies in order to expand their global presence. They would like to have better access to emerging biomedical markets in Asia. Because of the small size of some of the components needed in their products, they need micromolding manufacturing capabilities that exist in the United States and in Japan. Specifically, they have identified two small manufacturing companies, USMold and NipponMold.

USMold has historically been an exclusive provider to a large company in the United States. They are located in a small town in Texas, and have been run as a family business for 15 years. They take pride in the quality of their product

TABLE 3.4

Hofstede's Cultural Dimensions

Country	PDI	IDV	MAS	UAI	LTO
Arab World[a]	80	38	52	68	
Argentina	49	46	56	86	
Australia	36	90	61	51	31
Austria	11	55	79	70	
Belgium	65	75	54	94	
Brazil	69	38	49	76	65
Canada	39	80	52	48	23
Chile	63	23	28	86	
China[b]	80	20	66	30	118
Colombia	67	13	64	80	
Costa Rica	35	15	21	86	
Czech Republic[b]	57	58	57	74	13
Denmark	18	74	16	23	
East Africa[a]	64	27	41	52	25
Ecuador	78	8	63	67	
El Salvador	66	19	40	94	
Finland	33	63	26	59	
France	68	71	43	86	
Germany	35	67	66	65	31
Guatemala	95	6	37	101	
Hong Kong	68	25	57	29	96
India	77	48	56	40	61
Indonesia	78	14	46	48	
Iran	58	41	43	59	
Ireland	28	70	68	35	
Israel	13	54	47	81	
Italy	50	76	70	75	
Jamaica	45	39	68	13	
Japan	54	46	95	92	80
Malaysia	104	26	50	36	
Mexico	81	30	69	82	
Netherlands	38	80	14	53	44
New Zealand	22	79	58	49	30
Norway	31	69	8	50	20
Pakistan	55	14	50	70	0
Panama	95	11	44	86	
Peru	64	16	42	87	
Philippines	94	32	64	44	19
Poland[b]	68	60	64	93	32
Portugal	63	27	31	104	

TABLE 3.4 (*Continued*)

Hofstede's Cultural Dimensions

Country	PDI	IDV	MAS	UAI	LTO
Romania[b]	90	30	42	90	
Russia[b]	93	39	36	95	
Singapore	74	20	48	8	48
South Africa	49	65	63	49	
South Korea	60	18	39	85	75
Spain	57	51	42	86	
Sweden	31	71	5	29	33
Switzerland	34	68	70	58	
Taiwan	58	17	45	69	87
Thailand	64	20	34	64	56
Turkey	66	37	45	85	
United Kingdom	35	89	66	35	25
United States	40	91	62	46	29
Uruguay	61	36	38	100	
Venezuela	81	12	73	76	
Vietnam[b]	70	20	40	30	80
West Africa[a]	77	20	46	54	16

Source: Selected data from http://www.geert-hofstede.com (extracted January 25, 2009).

[a] Regional estimated values.

[b] Estimated values.

Note: Arab World = Egypt, Iraq, Kuwait, Lebanon, Libya, Saudi Arabia, United Arab Emirates; East Africa = Ethiopia, Kenya, Tanzania, Zambia; West Africa = Ghana, Nigeria, Sierra Leone.

Key: PDI, power distance index; IDV, individualism; MAS, masculinity; UAI, uncertainty avoidance index; LTO, long-term orientation.

and their sense of "Made in the USA." All of their management and operators are highly skilled and educated, and there is little diversity in the workforce.

NiponMold has customers in Asia, Europe, and America. Upper management is mostly Japanese; however, their technical workforce (e.g., senior engineers and machine operators) is composed of world-class specialists in their fields who have been recruited from countries like Germany, Sweden, Spain, and the United States.

Since the main target is the emerging Asian market, Japan appears to be better located than the United States; however, the cutting-edge biomedical technology in the United States makes it an attractive alternative. Also, some advantages of exchanges between Mexico and the United States may result from North American Free Trade Agreement's (NAFTA) regulations. Labor in the United States is more expensive than in Japan, and the main suppliers of raw materials are located in the United States. The products under

TABLE 3.5

Corruption Perceptions Index 2006

Rank	Country	Score	Rank	Country	Score	Rank	Country	Score
1	Finland	9.6	55	Namibia	4.1	111	Albania	2.6
1	Iceland	9.6	57	Bulgaria	4	111	Guatemala	2.6
1	New Zealand	9.6	57	El Salvador	4	111	Kazakhstan	2.6
4	Denmark	9.5	59	Colombia	3.9	111	Laos	2.6
5	Singapore	9.4	60	Turkey	3.8	111	Nicaragua	2.6
6	Sweden	9.2	61	Jamaica	3.7	111	Paraguay	2.6
7	Switzerland	9.1	61	Poland	3.7	111	Timor-Leste	2.6
8	Norway	8.8	63	Lebanon	3.6	111	Vietnam	2.6
9	Australia	8.7	63	Seychelles	3.6	111	Yemen	2.6
9	Netherlands	8.7	63	Thailand	3.6	111	Zambia	2.6
11	Austria	8.6	66	Belize	3.5	121	Benin	2.5
11	Luxembourg	8.6	66	Cuba	3.5	121	Gambia	2.5
11	United Kingdom	8.6	66	Grenada	3.5	121	Guyana	2.5
14	Canada	8.5	69	Croatia	3.4	121	Honduras	2.5
15	Hong Kong	8.3	70	Brazil	3.3	121	Nepal	2.5
16	Germany	8	70	China	3.3	121	Philippines	2.5
17	Japan	7.6	70	Egypt	3.3	121	Russia	2.5
18	France	7.4	70	Ghana	3.3	121	Rwanda	2.5
18	Ireland	7.4	70	India	3.3	121	Swaziland	2.5
20	Belgium	7.3	70	Mexico	3.3	130	Azerbaijan	2.4
20	Chile	7.3	70	Peru	3.3	130	Burundi	2.4
20	United States	7.3	70	Saudi Arabia	3.3	130	Central African Republic	2.4
23	Spain	6.8	70	Senegal	3.3	130	Ethiopia	2.4
24	Barbados	6.7	79	Burkina Faso	3.2	130	Indonesia	2.4
24	Estonia	6.7	79	Lesotho	3.2	130	Papua New Guinea	2.4
26	Macao	6.6	79	Moldova	3.2	130	Togo	2.4
26	Portugal	6.6	79	Morocco	3.2	130	Zimbabwe	2.4
28	Malta	6.4	79	Trinidad and Tobago	3.2	138	Cameroon	2.3
28	Slovenia	6.4	84	Algeria	3.1	138	Ecuador	2.3
28	Uruguay	6.4	84	Madagascar	3.1	138	Niger	2.3
31	United Arab Emirates	6.2	84	Mauritania	3.1	138	Venezuela	2.3
32	Bhutan	6	84	Panama	3.1	142	Angola	2.2
32	Qatar	6	84	Romania	3.1	142	Congo, Republic of	2.2
34	Israel	5.9	84	Sri Lanka	3.1	142	Kenya	2.2
34	Taiwan	5.9	90	Gabon	3	142	Kyrgyzstan	2.2

TABLE 3.5 (*Continued*)

Corruption Perceptions Index 2006

Rank	Country	Score	Rank	Country	Score	Rank	Country	Score
36	Bahrain	5.7	90	Serbia	3	142	Nigeria	2.2
37	Botswana	5.6	90	Suriname	3	142	Pakistan	2.2
37	Cyprus	5.6	93	Argentina	2.9	142	Sierra Leone	2.2
39	Oman	5.4	93	Armenia	2.9	142	Tajikistan	2.2
40	Jordan	5.3	93	Bosnia and Herzegovina	2.9	142	Turkmenistan	2.2
41	Hungary	5.2	93	Eritrea	2.9	151	Belarus	2.1
42	Mauritius	5.1	93	Syria	2.9	151	Cambodia	2.1
42	South Korea	5.1	93	Tanzania	2.9	151	Cote d'Ivoire	2.1
44	Malaysia	5	99	Dominican Republic	2.8	151	Equatorial Guinea	2.1
45	Italy	4.9	99	Georgia	2.8	151	Uzbekistan	2.1
46	Czech Republic	4.8	99	Mali	2.8	156	Bangladesh	2
46	Kuwait	4.8	99	Mongolia	2.8	156	Chad	2
46	Lithuania	4.8	99	Mozambique	2.8	156	Congo, Democratic Republic of	2
49	Latvia	4.7	99	Ukraine	2.8	156	Sudan	2
49	Slovakia	4.7	105	Bolivia	2.7	160	Guinea	1.9
51	South Africa	4.6	105	Iran	2.7	160	Iraq	1.9
51	Tunisia	4.6	105	Libya	2.7	160	Myanmar	1.9
53	Dominica	4.5	105	Macedonia	2.7	163	Haiti	1.8
54	Greece	4.4	105	Malawi	2.7			
55	Costa Rica	4.1	105	Uganda	2.7			

Source: Adapted from Transparency International, *Annual Report 2006.*

Note: The score relates to perceptions of the degree of corruption as seen by businesspeople and country analysts, and ranges between 10 (highly clean) and 0 (highly corrupt).

consideration are very small and most likely will be transported to the customers via air.

The Decision Problem

Assume you are hired by MexPlast to help them decide whether to partner with USMold or NipponMold. MexPlast is mainly interested in an analysis that considers the globalization perspective.

This problem can be set up as a hierarchical decision problem, as illustrated in Figure 3.4.

To start calculating the weights shown in the decision tree, it is necessary to set up the matrices for the pair-wise comparison between the criteria under consideration, namely, culture and socioeconomic factors. Assuming that

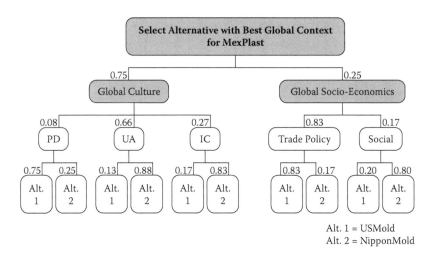

FIGURE 3.4
Decision tree for selecting alternatives.

TABLE 3.6

Comparison and Normalized Matrices between Culture and Socioeconomics

Criteria	Cult	Soc/Eco	Criteria	Cult	Soc/Eco	Weight
Cult	1	3	Cult	0.75	0.75	0.75
Soc/eco	1/3	1	Soc/eco	0.25	0.25	0.25
	1.33	4				

you consider that culture is somewhat more important than socioeconomics, you assign culture a 3 in the preference scale. The initial and normalized comparison matrices are as shown in Table 3.6.

Then we consider the next hierarchy level: culture and socioeconomics. Note that the *culture* criterion has several dimensions. For this example we will consider three of Hofstede's cultural dimensions: power distance (PD), uncertainty avoidance (UA), and individualism (IC). The next AHP step is for the decision maker to rank the importance of each of these dimensions with respect to the culture criterion. Assume that you more than strongly put more importance on uncertainty avoidance than on power distance, and put somewhat more importance on uncertainty avoidance than individualism. Also, you strongly prefer individualism over power distance. The comparison matrix could look as in Table 3.7.

Similarly, we assume that the *socioeconomic* criterion is composed of two factors: trade policies and social stability. Further, assume you strongly put more importance in trade policies than social stability. The relative importance between trade policies and social stability with respect to the socioeconomic criterion could be calculated as shown in Table 3.8.

TABLE 3.7

Comparison and Normalized Matrices among the Cultural Dimensions

Cult	PD	UA	IC	Cult	PD	UA	IC	Weight
PD	1	1/7	¼	PD	0.08	0.10	0.06	0.08
UA	7	1	3	UA	0.58	0.68	0.71	0.66
IC	4	1/3	1	IC	0.33	0.23	0.24	0.27
	12	1.48	4.25					

TABLE 3.8

Comparison and Normalized Matrices between Socioeconomical Dimensions

Soc/Eco	Trade	Social	Soc/Eco	Trade	Social	Weight
Trade	1	5	Trade	0.83	0.83	0.83
Social	1/5	1	Social	0.17	0.17	0.17
	1.20	6				

We now consider the last hierarchy level, which consists of comparing USMold and NipponMold with respect to each of the criteria (or factors) in the previous level, namely, PD, UA, IC, trade, and social. In this example we assume the decision maker's preferences are summarized in the comparison matrices from Table 3.9.

These weights are shown by the corresponding box on the decision tree in Figure 3.4.

We can now compute the ranking of USMold and NipponMold:

$$Rank(Alt\ 1) = 0.75[0.08(0.75) + 0.66(0.13) + 0.27(0.17)] + 0.25[0.83(0.83) + 0.17(0.20)] = 0.32$$

$$Rank(Alt\ 2) = 0.75[0.08(0.25) + 0.66(0.88) + 0.27(0.83)] + 0.25[0.83(0.17) + 0.17(0.80)] = 0.69$$

Before using these ranks we verify the consistency of the comparison matrices. All the 2×2 matrices are perfectly consistent. We only need to analyze the 3×3 matrix (culture):

$$n_{max} = 0.08(1 + 7 + 4) + 0.66(1/7 + 1 + 1/3) + 0.27(1/4 + 3 + 1)$$
$$= 0.08(12) + 0.66(1.48) + 0.27(4.25) = 0.96 + 0.97 + 1.15 = 3.08$$

$$CR = \frac{n(n_{max} - n)}{1.98(n-1)(n-2)} = 3(0.08)/[1.98(2)(1)] = 0.24/3.96 = 0.06 < 0.1$$

All the comparison matrices are consistent, so we can conclude that the rankings obtained are reliable; i.e., NipponMold is more attractive than USMold to deal with MexPlast considering the globalization perspective.

TABLE 3.9

Comparison and Normalized Matrices between the Two Alternatives with Respect to Each Decision Criterion

PD	Alt 1	Alt 2	PD	Alt 1	Alt 2	Weight
Alt 1	1	3	Alt 1	0.75	0.75	0.75
Alt 2	1/3	1	Alt 2	0.25	0.25	0.25
	1.33	4				

UA	Alt 1	Alt 2	UA	Alt 1	Alt 2	Weight
Alt 1	1	1/7	Alt 1	0.13	0.13	0.13
Alt 2	7	1	Alt 2	0.88	0.88	0.88
	8	1.14				

IC	Alt 1	Alt 2	IC	Alt 1	Alt 2	Weight
Alt 1	1	1/5	Alt 1	0.17	0.17	0.17
Alt 2	5	1	Alt 2	0.83	0.83	0.83
	6	1.2				

Trade	Alt 1	Alt 2	Trade	Alt 1	Alt 2	Weight
Alt 1	1	5	Alt 1	0.83	0.83	0.83
Alt 2	1/5	1	Alt 2	0.17	0.17	0.17
	1.2	6				

Social	Alt 1	Alt 2	Social	Alt 1	Alt 2	Weight
Alt 1	1	¼	Alt 1	0.20	0.20	0.20
Alt 2	4	1	Alt 2	0.80	0.80	0.80
	5	1.25				

Key: Alt 1, USMold; Alt 2, NipponMold.

Conclusions

This chapter described a method to aid engineers in making decisions considering both qualitative and quantitative information. The method consists in integrating the global engineering model (GEM) presented in Chapter 2 with a well-known decision methodology called analytical hierarchical process (AHP). The application of the methodology was illustrated through a detailed development of two examples.

Review and Study Questions

1. Consider the decision of choosing between two companies (A and B) that have offered you a job. Assume that you would like to evaluate this decision giving consideration to the following three criteria: salary, location, and professional growth potential. You have gathered the following information:

   - Assume that you have the same preference level for salary and professional growth, but that salary is more than important for you than location.
   - Annual salary: Company A is offering you $65,000 and Company B is offering you $55,000.
   - Location: In both cases you will need to relocate; however, you strongly prefer the location offered by Company B.
   - Professional growth: Both companies offer similar opportunities for professional growth.

   Answer the following questions:

   a. Draw the corresponding AHP decision tree.

   b. You are asked to rank alternatives A and B using AHP. Clearly define the necessary comparison matrices, including a brief explanation of the values used in each matrix. Write the relative weights obtained on the tree from part (a).

   c. Also determine and comment on the consistency of the comparison matrices used in part (b). The example in the beginning of the chapter yielded an inconsistent comparison matrix for the location's stability criterion. Rework the problem by using a preference of 3 instead of 6 when describing the preference of City C over City B in the comparisons with regard to stability. Explain why this change reflects a more consistent judgment.

2. A company is considering setting global operations in Hong Kong (HK), Mexico (MX), or Israel (IS). Management realizes that national work culture must be considered in the selection of the country for its new operations. Management would like to quantify the degree to which the cultures of the candidate countries are similar to U.S. culture. Assume that you will use Hofstede's cultural dimensions— power distance (PD), uncertainty avoidance (UA), and individualism (IC)—as the main criteria for comparison. You must give each criterion the same relative weighting in your ranking. Answer the following questions:

   a. Develop a ranking of these three countries (use the comparison matrices given below). Write the partial weights on the AHP tree shown below.

   b. Check the consistency of the comparison matrices used. Comment on your results and analyze inconsistencies if found.

3. Consider the "MexPlast Goes Global" case study and Hofstede's PD, UA, and IC indices given in the chapter. Do you agree with the preference values given in the matrix comparing USMold and NipponMold with respect to each cultural dimension given in the case study? Explain.

Notes

1. Thomas L. Saaty, *Fundamentals of Decision Making and Priority Theory with the Analytic Hierarchy Process* (Pittsburgh, PA: RWS Publications, 1994).
2. V. Jorge Leon, "Operations Research in Manufacturing," in *Manufacturing Engineering Handbook*, ed. H. Geng (New York: McGraw-Hill, 2004), chapter 15.
3. Geert Hofstede, *Culture's Consequences*, 2nd ed. (London: Sage, 2001).

Section II

Case Studies:
Cultural Emphasis

4

MEPO Manages Culture

When people think about multinational companies operating in Mexico, they often assume that the organization is based in the United States and operates near the U.S.-Mexico border in *maquiladoras*. It is true that the United States is Mexico's largest trading partner, and largely because of the North American Free Trade Agreement (NAFTA) U.S. investment in Mexico has grown rapidly during recent years. However, firms from many countries operate within Mexico, and have done so for decades. In 2000 the European Union signed a free trade agreement with Mexico, stimulating even more interest in the Mexican market among European firms. The largest economies in Europe—Germany and the United Kingdom—have the largest presence in Mexico, but Spain and France also are important trading partners. Currently, the total annual flow of investments controlled by French interests in Mexico is approximately US$600 million, and the value of the products produced by French firms exceeds $8 billion. French firms employ 70,000 Mexican workers.[1]

French and Mexican Culture: The Best of Both Worlds?

On the surface, French and Mexican cultures seem to be quite similar. As Chapter 1 pointed out, the most important factor influencing cross-cultural interaction seems to be uncertainty avoidance (UA). UA refers to the extent to which people feel threatened by new, uncertain, or ambiguous situations. Of the fifty-three countries that Hofstede originally studied, France and Mexico had among the highest UA scores (86 and 82, respectively). Only ten countries had higher UA scores than France, and only seventeen had higher scores than Mexico. (As a basis of comparison, the U.S. score was 46, for a rank of 43 out of 53.) This similarity was a little surprising, because UA scores tend to be low in countries like France that are relatively wealthy and have long-term, stable political systems. However, UA is a description of how people perceive and feel, and people can feel insecure even when the political and economic situation surrounding them is stable.

Workers in high-UA cultures tend to feel more anxious than those in low-UA cultures, and to be pessimistic about the motives underlying their organizations' actions, especially over issues involving employees' welfare. This pessimism can be offset by the sense of security that comes from having long-term relationships with their organizations, especially if those organizations are large enough, competitive enough, and their decision makers competent enough to make it seem to be a "safe bet" for the future. Their anxiety can be reduced by the existence of clear organizational rules that they follow and can realistically expect others to follow.[2] Both French and Mexican workers expressed a high level of faith in technical expertise and knowledge, and believed that specialized tasks and training are important means of ensuring that workers will be experts in their duties. Other researchers have found that "knowing it all" is very important to French managers and specialists. As a result, subordinates would never challenge the expertise or judgment of their supervisors, and rarely take action without obtaining their supervisors' permission. Ironically, this process seems to open up communication between French supervisors and subordinates, because both groups are constantly checking back with one another. It also helps maintain an image of supervisors' expertise because their subordinates are constantly keeping them informed and sharing their own expertise in the process.[3] Mexican workers obtain a degree of security through their interpersonal relationships, with family, friends, and co-workers. In collectivist societies like Mexico, the family is the most important organization, and its importance is always reflected in all social aspects (i.e., in school, work, friends, etc.). Mexicans are most comfortable when surrounded by familiar others, in order to talk and to share their own experiences. Forming relational networks creates stability zones, which allow them to work more easily with members of out-groups, foreigners, or other people they do not know personally. In organizations, working with out-group members eventually allows them to develop the level of trust necessary to work comfortably with outsiders, but initially these relationships are likely to be strained.

French and Mexican cultures also seem to be similar on Hofstede's second most important dimension, power distance (PD), the extent to which people expect power, wealth, and privilege to be distributed unequally among members of a society or organization, accept those inequalities, and respond to them. In Hofstede's original research, French employees had high PD scores (an average of 68, for a rank of 15 out of 53), and those of Mexican employees were even higher (an average of 81, for a rank of 5). In a high-PD society, inequality is viewed as inevitable—superiors are superiors because they are superior persons—and good, because it is the basis of a stable and ordered society.[4] "Acting properly," which means acting according to the society's rules for one's social standing or organizational rank, is of primary importance. Indeed, it is even more important than efficient performance of one's tasks. Organizations in high-PD cultures tend to be very bureaucratic, with power and decision making centralized at the top of the chain of command. Subordinates do not feel free to disagree with their supervisors, and tend to prefer supervisors who use autocratic styles of leadership. Similarly, supervisors prefer—and use—autocratic leadership strategies. When things go wrong, low-power/status people are blamed. On the surface, these assumptions seem to make life exceptionally easy for superiors. However, over the long term, high levels of subservience and dependency create a latent conflict between the powerful and powerless. When a situation becomes intolerable, the only solution is to change the persons who are in charge—a potential for revolt is built into the system of deference.[5]

The two societies do seem to differ in terms of the ways in which PD is implemented. In highly communitarian societies like Mexico, power distance seems to operate within referent groups. This means that people defer to superiors with whom they have significant relationships—the patriarch of a family, the supervisor of a work group that has been in existence for a long time, and so on. The deference does not extend to people outside of key referent groups. This is why Mexican workers are deferential to superiors, while strongly disliking working for a supervisor of a different nationality or race. Over time, and with a great deal of interaction, outsiders can be accepted into an in-group, and begin to experience the advantages of holding a superior position. But as long as they remain outsiders, power is withheld.

Conversely, in a highly individualistic culture like France, power distance functions very differently. For example, of the nine European countries studied by Barsoux and Lawrence, France was next to last in the percent of workers who agreed that they would "carry out instructions from my supervisor," but first in the percent who said they would do so if they were persuaded that their supervisor was right. But, France also is a high-PD culture, which means that it is very unlikely that a subordinate would disagree with his or her supervisor in public or overtly refuse to follow orders. Some observers attribute this behavior to a value of politeness, but it is more complicated than that term suggests. Sociologist Paul d'Iribarne concludes that the French manage the dilemma created by being both individualistic and deferential through an honor principle and bureaucratic structure. As long as a supervisor gives orders that are appropriate to his or her class and position in the formal structure of the organization, subordinates will follow them, especially if they are presented as impersonal, bureaucratic demands rather than direct orders from one person to another. Michel Crozier sums up this honor principle in his classical study of French culture:

> Face-to-face dependence relationships are ... perceived as difficult to bear in the French cultural setting. Yet, the prevailing view of authority is still that of universalism and absolutism ... the two attitudes are contradictory. However, they can be reconciled within a bureaucratic system since impersonal rules and centralization make it possible to reconcile an absolutist conception of authority and the elimination of most direct dependence relationships.[6]

By giving only appropriate orders, supervisors allow their subordinates to retain their honor, and by following those orders, subordinates protect the honor of their supervisors.

In sum, French and Mexican cultures coincide in important ways. Both place a great deal of faith in specialized expertise and support deferential relationships between subordinates and superiors, although superiority is defined in terms of class for the French and in terms of role in relational groups in Mexico. Both cultures support a degree of suspicion about the motives of organizations, especially about the welfare of their workforce, but that suspicion declines in long-term organization–employee relationships. High power distance implies that supervisors are granted a great deal of deference, but if they try to exercise their power outside of the work situation, or if they make unwise decisions or otherwise fail in their duties, that deference will rapidly disappear. Supervising workers will be most difficult for outsiders who have not yet established personal relationships with their subordinates, especially if they do things that accentuate their difference.

The Facts of the Case

MEPO (name changed for confidentiality reasons) is a French consortium dedicated to the fabrication of plastic products. MEPO has had extensive experience operating in other countries, and currently has more than thirty-five plants in nineteen countries on four continents. MEPO installed its first plant in Mexico in 1997. The plant that is examined in this case study is dedicated to producing automotive parts such as fenders, defenses, air grills, etc. It sells most of its products to a nearby German automobile assembly plant that has had great success in the Mexican market with one of its recent models. It has a smaller contract, negotiated under the EU-Mexican free trade agreement, through which it provides similar parts to a Japanese-French automobile manufacturer. The MEPO plant operates on a just-in-time inventory control process. MEPO's high-level operations, research and development (R&D), engineering analysis, and product design are located in France. Worldwide operations of MEPO are shown in Figure 4.1.

MEPO's plant employs two hundred workers, most of whom are Mexican. Most of the workers are trained in Mexico. However, MEPO is interested in providing advanced training to its international specialists. It has sent several technicians and engineers to French schools for advanced training, and has rotated some of its Mexican personnel to other MEPO plants around the world so that they can learn particular techniques. When the plant encounters specific production or technical problems, a French engineer who is an expert in the area is dispatched to Mexico. They stay for only short periods of time and focus only on the problem they were sent to solve.

The plant is highly mechanized, including machinery for automatic injection processes and robotized painting. Beginning in January 2002, the plant steadily increased its production of fenders for the French-Japanese company

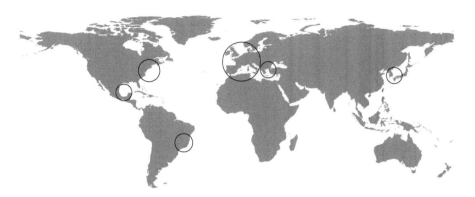

FIGURE 4.1
MEPO's worldwide operations.

from sixty-four parts per shift to four hundred parts per shift in May of the same year. As the pace of production increased, the number of defects also increased. It was at this point that our research team was invited to begin our observations of the plant. Over a 6-month period we distributed question- naires among the employees and analyzed their results, engaged in direct observations of the employees in action, and interviewed plant management, production supervisors, and workers.

Problem 1: What's Wrong with the Molds?

The parts that came out of MEPO's plastic injection process yielded smaller dimensions than required. The injection combined with a baking process that was conducted when the parts reached the customer produced some parts that were 1 or 2 mm below the required specification dimensions, which cre- ated excessive looseness in assembly with the adjacent parts. For the Mexican engineers and managers, the defects simply did not make sense. MEPO uses identical molds and equipment in plants located in France, Turkey, Spain, and Brazil. Only in the Mexican plant did the parts fail to reach the design dimensions. In line with MEPO's standard procedures, a French technician who was an expert in this particular equipment was dispatched to Mexico to solve this problem. Over the years, Mexican technicians and workers had become accustomed to having French specialists visit the plant on problem- solving assignments. They had learned to trust the specialists very much. But, this one was different. He refused to listen to the Mexican workers, regardless of their tasks or level of education. He evidently believed that he was the expert on the very specialized equipment that was malfunctioning, and no one else's opinion or expertise mattered. In a sense, he was correct, because even the Mexican engineers had a much more generalized training than he did, and much less expertise on the specific equipment that was mal- functioning. Without anyone else's input, he decided to increase the quantity of the material being injected into the molds and to increase the pressure with which it was injected. Not only did these steps not solve the problem, but they also generated excessive burrs and misadjusted the alignment of the molds. As a direct consequence of these steps, production was delayed and production costs skyrocketed.

A second French expert was called to solve the problems. He too was an expert in the injection system, but he had experience troubleshooting in MEPO plants around the world. Instead of taking the problem on alone, he started by making an exhaustive survey of the opinions and expertise of the Mexican workers, technicians, and engineers. He then created a problem-solving team, composed of himself and the workers, technicians, and engineers who were most directly involved with the equipment. At some point in one of these meetings—no one seems to remember who made the discovery or how it was made—someone noted the obvious. Unlike any other MEPO plant using this technology, the Mexican plant is located 2,100 meters above sea level with an average annual temperature of 23°C. In order to compensate for the altitude and temperature, the equipment needed to be operated within different parameters—molding temperature, injection speeds, percentages of injected plastic mass, and retention time all had to be revised. In addition, when these parameters were changed, they changed the behavior of the materials in secondary ways. For example, part shrinkage is actually lower at higher molding temperatures. Higher temperatures lower the viscosity of the polymer, allowing more material to be packed into the mold. The effect of the last is the same as using higher injection pressures. If all of these changes were made, the parts could finally achieve the specified dimensions. Of course, the irony is that it took no specialized expertise to realize that altitude and temperature influence the behavior of polymers. It did take special skill to determine how to compensate for the environmental factors, so it was good to have the French engineer on site.[7]

Problem 2: How Do We Make That Fender Fit?

Figure 4.2 shows the fender subassembly.

Once a fender leaves the injection press, six separators and three fasteners are attached to it, totaling nine additional components in the fender subassembly. However, MEPO's Japanese-French customer required the company to enlarge one of the holes through which the fender was fastened to the car chassis. Eventually, MEPO's technicians demonstrated that the fender could be perfectly assembled without reworking the hole in question. However, the customer firm is very bureaucratic and the local supervisor had to request authorization from his home office before he could permit the change in design. This further delayed the project, but thanks to the successful negotiation, MEPO now saves in labor costs.

In addition, there were problems in the subassembly process. MEPO's workers often made mistakes installing the required separators and

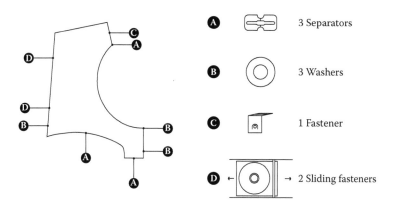

FIGURE 4.2
The fender subassembly.

fasteners to the plastic fenders. As a result, MEPO's line managers had to spend a great deal of time and effort inspecting the production and adding the missing components. This problem may have resulted from delays in the workers' training in the use of the new technologies; in fact, production using the technology started while the workers were still taking quality control courses. The line managers' temporary solution was to involve themselves in the fender subassembly inspection process. This decision is consistent with Mexican paternalism, which posits that supervisors are responsible for protecting their subordinates from punishment until they can correct their errors, and makes it difficult to completely delegate responsibilities to subordinates.[8] But, it reduces the quality of life of the line managers because it makes them finish their work late. While this, too, is considered to be very natural in Mexico, it can hide production problems that would best be corrected quickly in order to maintain maximum efficiency.

Finally, MEPO encountered difficulties regarding quality control of the fender production process. The company had established a quality assurance program for the fenders that involved taking five fender samples of each injection run in order to verify the accuracy of the various dimensions. The sample parts are mounted in a fixture, and MEPO operators proceed to measure the coordinates of twenty-two different points along the fender with a coordinate measuring machine (CMM). A CMM allows workers to take accurate three-dimensional measurements of an object. This level of testing was consistent with MEPO's own internal standards and with the desires of its German customer. However, MEPO's Japanese-French customer only requires dimensional measurements to be taken at nine checkup points. When its engineers received the test results, they were confused by the additional measurements shown on the MEPO drawings. In addition, the extra measurements increased the costs involved in the quality assurance system beyond the parameters that the French-Japanese buyer had expected. Eventually, MEPO was able to explain the differences to both customers and negotiate acceptable cost sharing, but in the short term it created a good bit of confusion and time-consuming communication.

Eventually the confusion was prevented by shifting to a different system of quality assurance. Today the fender subassembly inspection is done utilizing a Poka-Yoke device. This term was coined by the creators of Japanese management techniques and has been widely adopted by advocates of Lean manufacturing. *Poka-Yoke* means "foolproof," and Poka-Yoke systems are designed to prevent production errors before they occur rather than detecting and correcting them after they happen. For example, if someone has to assemble two wings to a toy airplane, the joints are designed so that it is impossible to get them reversed—one connection is rectangular and the other is oblong, for example. However, the system is foolproof only if the operators understand how to use it. In MEPO's case, the device was designed by managers and built outside the plant rather than by the workers who would use it or the local engineers and technicians who would be in charge of its use. They wrote an extended instruction manual (in Spanish) and made it available to the workers. However, the workers did not read it on their own, as the designers intended. They waited until their superiors came to them to personally explain its operation. Of course, more delays and more production errors took place in the interim. As important, MEPO obtains a great deal of its overall organizational efficiency by requiring all of its plants to manufacture parts in such a way that a plant can complement the production of another plant in the event of capacity shortage or maintenance. Whenever one plant makes adjustments to accommodate one customer, it reduces the advantages of substitutability.

Interpretation

Figure 4.3 refers to the center ring in our global engineering model (GEM), in which we will first interpret this case study.

In Chapter 1 we explained that the core concept needed to understand engineering in multinational corporations is that of the global organization. It has two dimensions, the mindset of a corporation and its leadership, and the structures that had been put in place to help an organization implement that mindset. The global mindset of MEPO's management is a positive one for cross-national operations, one that is likely to enhance intercultural operations and minimize cross-cultural misunderstandings and conflict. This favorable global organization was in place from the first days of the project.

At least in terms of cultural coordination, the optimal way to start a business in another country is through a greenfield start, in which the multinational

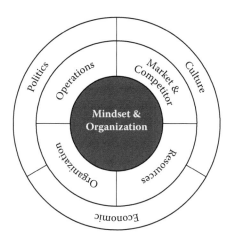

FIGURE 4.3
Global engineering model's center ring influence.

firm sets up a foreign subsidiary from scratch by sending a single expatriate or a small group to start and grow a local business. Greenfield starts take a great deal of time, but during that time the expatriate team can learn a great deal about the local culture, and indigenous employees can become accustomed to the company and its employees. The team also can get to know local employees well enough to know who to hire so that they will adapt most easily to the culture of the overall organization. Greenfield starts have very high success rates, especially in high-uncertainty-avoidance cultures like Mexico. The changes created by the new company are minimal, and local people have an opportunity to bring at least some of the expatriates or the local people hired by the company into their in-groups. Uncertainty, and thus resistance, is minimized.[9] Because of the time and expense involved, multinationals tend to use greenfield strategies only when their managements trust the competence and motivations of local people, hope to build a long-term, equitable relationship with them, feel comfortable with operating in other countries, and understand the local culture.

MEPO's greenfield start had all of the key uncertainty-reducing characteristics. It was clear to everyone concerned that MEPO's management wanted to build a long-term working relationship with local peoples and with the other multinational corporations operating in the area. This mindset was implemented through a distinctive organizational structure. The plant was operated almost completely by Mexicans, and the home office gave them a great deal of autonomy in running its day-to-day activities. Even the visiting specialist system was structured in a way that served to minimize cultural disruption. Specialists were called in to deal with specific, clearly defined problems, and there was a clear expectation that they would return to France once their task was completed. The plant was in constant contact with the home office, but it was not being overtly watched or controlled by France. As a result, even when a specialist behaved in culturally inappropriate ways, as the first engineer sent to deal with the altitude problem did, the relationship between the plant and the home office was strong enough to avoid any significant damage.

Even in the best of greenfield startups there still are challenges involved in integrating two groups of employees. French engineers are trained as specialists, and hold engineering generalists, like Mexican engineers, in lower regard. They also learn a different form of problem solving. Systems are planned and implemented according to established blueprints, only after necessary resources are obtained. For them, the form of decision making is as important as the outcome. In contrast, Mexican engineers learn to devise creative solutions on an *ad hoc* basis, as a solution to specific problems as they crop up. If a strategy works, that is all that's important. As a result, the French engineers did not value or trust recommendations made by their Mexican peers, and Mexican engineers became frustrated with the French insistence on logical and structured planning. The French employees also were concerned about maintaining control of their Mexican subsidiary, and sometimes did so by refusing to share information with the Mexican workers. For example, all technical information and decisions were given in French in order to control access to key information. After 6 years of experience working together, these problems have become less pronounced. But, they indicate that even the best-intended mindsets and best-designed structures are difficult to implement.

There were, however, other elements of the structure that created long-term problems. The first involves organizational learning. As far as our research team could tell, MEPO has no system in place through which the insights that a specialist learns during one of his or her visits to a plant or office can be communicated to other specialists in the home office, or even stored in an archive that can be consulted when other specialists confront similar situations in the future. Although it might seem obvious that every company that uses troubleshooters should have a system like this in place, they are surprisingly rare. For example, Diane Vaughan's extensive analysis of the space shuttle *Challenger* accident found similar problems at NASA. Each shuttle launch was in some ways unique—different goals, payloads, duration of mission, and so on. They also are amazingly complex. Consequently, NASA created a matrix structure in which a team is composed of an expert in every area that is relevant to a particular mission. At one point, there also was a complicated structure in place to disseminate information about the technical problems and solutions encountered in each mission throughout the organization. But, over time and with a series of budget cuts, these systems were eroded. As a result, once a mission was over, its team compiled the information they learned, and stored it away, often never to be consulted again. Information about problems with the solid-fuel booster rockets' O-ring seals, present in many missions before the *Challenger* launch, was one of the items that was filed away.[10] Volkswagen, which uses a consulting specialist system much like MEPO's in its offshore plants, recently implemented an information sharing system after recognizing that it too had no effective structure for ensuring that the organization as a whole would learn from these interventions. There was no way for our research team to determine why MEPO has not implemented such a system. It may be that the corporate culture defines problem solving as the solving of *individual* problems, not as a means of organizational learning. Or it may result from the organizational structure in which each offshore plant reports directly to headquarters, rather than to one another. Such a structure may make it difficult to transfer technical information from plant to plant, short-circuiting that learning process. Whatever the reason, it is clear that the structure does not encourage widespread learning.

The final aspect of global organizational structure involves the consortium itself. In short, MEPO has multiple masters—the home office, of course, but also the companies to which it sells parts. Although fenders look like fenders, there were subtle differences in the requirements imposed on the plant by the home office, its French-Japanese buyer, and its German buyer. Running separate fabrication processes for each company would have been prohibitively expensive for MEPO. But, trying to accommodate very different sets of specifications with the same process created some very serious headaches. Unfortunately, these problems seem to be built into the nature of engineering in a global economy. The economic advantages of globalization depend on consortium-like organizational structures, but they make global engineering exceptionally difficult.

In Chapter 1 we also pointed out that not every level of our model of global engineering would be important to every case. MEPO's Mexican operation was so well designed and so well operated that the three considerations we included in our center ring—global operations and supply chain, global market participation, and global resources—were relatively unimportant. This is likely to be the case whenever a multinational corporation enters a different nation in order to establish a long-term presence in the local or nearby markets. Supply chains are likely to be relatively short, the market is close at hand and understood by the local managers and the central office, and the resources—both physical and intellectual—are either close by or easy to transport from the home office. However, the outer ring, especially the cultural dimension, was salient (see Figure 4.4).

French and Mexican cultures are similar in very important ways. But, they are not identical, and the most important difference—the value placed on individualism and community—created a possibility of culture clash. The long-term relationships that MEPO had built with its Mexican employees helped minimize that potential. The first specialist's insistence on "going it alone," without consulting his Mexican counterparts, could have damaged the relationship between the company and its Mexican workforce. But, it did no lasting damage. He eventually went home, as everyone knew he would, and the local workers saw his extreme isolation and individualism

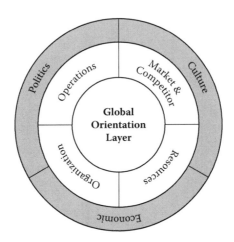

FIGURE 4.4
Global engineering model's outer ring influence.

as a personal characteristic, not as a comment on MEPO or French people in general. Replacing him with a more collaborative, culturally sensitive expert further reduced the negative impact of his visit. In high-power-distance cultures like Mexico's, lower-ranking people defer to higher-ranking ones. But, if they fail, they are responsible for that failure, and can and should be replaced. MEPO's specialist system did precisely what is valued in a high-PD culture: the authority figure failed and was replaced.

However, there are two important lessons about cross-cultural engineering that can be drawn from MEPO's experience. First, culture *influences* behavior but it does not *determine* it—when people from even very different cultures work together, culture clash is not inevitable, and when it does occur, it can be managed successfully. This is in part because cultural tendencies are variables, and as such are distributed normally within a population. As Chapter 1 pointed out, some cultures are more homogeneous than others, but even within the most homogeneous cultures there still is a degree of variability. In highly individualistic cultures, many people still have connections to important referent groups, and many of them are influenced more by their commitments to these groups than to their desire for individual rewards or gain. Conversely, even in highly communitarian cultures, individual needs and goals still are present, and may even be dominant for some people. In addition, people are not trapped within their own cultural frames of reference. They can learn to understand how the taken-for-granted assumptions of their cultures influence their attitudes, perceptions, and behaviors. They also can learn to understand how other people are influenced by their cultures. This learning may come informally, through experience with another culture, or it can be obtained through more formal training. The second specialist that MEPO brought to Mexico had learned these lessons; the first one had not.

The second culture-related lesson to be learned from MEPO's experience is that cultural differences can operate in very subtle ways. MEPO found a simple solution to the fender problem—the Poka-Yoke device, originally developed by the Toyota company. They adapted it and wrote a clear training manual for its use. But, the device was built outside of the plant, so the workers had no direct experience with it and no reason to feel committed to its use. In addition, the workers simply did not read the manual. Among developing countries, Mexico's literacy rate is quite good. But, Mexico is a very visual culture, so people learn to understand diagrams and pictures more readily than printed words. In addition, training works best in community-oriented cultures if it is done within in-groups and led by the ranking member of those groups.[11] Ironically, the Mexican engineers' expectation that workers would individually read the manual is much more appropriate for a French (or U.S.) workforce than for a Mexican one. It was not until superiors within the workers' referent groups began working with them that the training succeeded, as one would expect in a high-power-distance culture.

Conclusions

Every global manufacturing project has to be soundly based on technical as well as cultural analysis. Guidelines for multicultural work teams should not be generalized. Every culture has different characteristics, which can influence the work of the team in many different ways, and have significant effects on the success of the project. Through several discussions and agreements with MEPO managers, the research team offered a number of recommendations for global manufacturing collaboration. The following are some of the ideas proposed:

- When performing global manufacturing, educate and train the workforce not only in technical but also in cultural issues.
- Prepare brochures depicting the cultures of the people from the different countries represented in the company operations.
- Have the workers tell stories about their past experiences—positive and negative—working with people from the cultures they will be interacting with. Circulate the cases to all the workforce through culturally appropriate media (in writing for French employees and through in-group discussions among Mexican employees).
- Recognize that French and Spanish languages appear to be similar because of their Latin roots. However, pronunciation and grammar are completely different. Try to teach at least some of the other

country's language to all employees, and especially to the ranking employees within each work group. Most people appreciate when people from other countries or other cultures try to speak and understand their language. Of course, the more fluent employees are with both languages, the better. Also recognize that some terms are especially likely to create misunderstanding. Because Mexican and French cultures have different conceptions of time, words like *ahorita* (interpreted by French as meaning "right now") and *luego* (which means "after," but does not express a clear sense of how long after) are especially ambiguous. Compile a glossary of terms to avoid misunderstandings. Include technical and foreign words directly related to the production type. Provide similar instructions about nonverbal cues. For example, our team observed many cases in which the French tendency to talk loudly and move their arms and hands vigorously seemed impolite to Mexicans. Conversely, French workers sometimes failed to realize that a Mexican speaker was very emotionally involved in a discussion because he or she used a more subdued voice and gestures than an equally emotional French speaker would use.

- Exchange or send workers to other plants of the same company in different countries for technical as well as cultural training.
- Emphasize universal values, such as respect to other cultures. Disseminate the company's global mindset among all the workers (including its mission, vision, objectives, and beliefs). Teach work ethics courses and give advice and plan to have an experienced counselor.

More specific recommendations for Mexican workers in relation with French workers should include:

- Observe French punctuality rules. Mexicans being late for something is more objectionable to French workers than French workers being early is to Mexican employees.
- When communicating with French employees, provide as much information as possible. Mexicans are accustomed to coping with incomplete information because they are from a high-context culture, but French workers are uncomfortable with it.
- Recognize that personal relations with the French people will seem to be emotionally colder and more formal than is usual in Mexican-to-Mexican relations.
- Recognize the importance of agreements in French culture. French workers have learned to not work on a project unless a clear and detailed contract has been specified in advance. Changes in the contract must be negotiated openly. Realize that this will be frustrating

for Mexican workers who want to "get on with the job," and that it does not mean that their French counterparts do not trust them as individuals or that they do not trust Mexican people in general.

More specific recommendations for French people coming to Mexico are:

- Learn about Mexican culture before arriving on site. Study should focus on learning the attitudes and values in Mexican culture that can be used to improve work in multicultural teams. Try to understand the nuances of Mexican culture. For example, high PD means that Mexican workers readily accept orders from supervisors and easily change their behavior in order to obtain acceptance from higher-ranking people. But, this paternalistic style also places demands on the higher-ranking people—requirements to protect one's subordinates and do what is necessary to make the Mexican workers feel confident that recommendations are based on sound analysis and careful consideration of the problem at hand. Take time to observe and learn the culture behavior before starting to take actions.

- Adapt to the work team instead of forcing the team to adapt to French individualism. Gain the team's confidence based on respect and not by using authoritarianism and superiority demonstrations.

- Be aware that in spite of standardization of manufacturing practices, each culture has its own work personality.

We started this chapter by noting that similarities between French and Mexican cultures provide the basis for effective multinational operations. This is especially true with a company like MEPO, which has a truly global orientation, one that values long-term, equitable relationships between the home office and subsidiaries, and creates the structures necessary to implement that vision. However, subtle cultural differences, and the unintended consequences of policies, procedures, and structures, combined to make collaboration more difficult to sustain than an overview of the two cultures would have predicted. In the following chapter we profile another case, one in which cultural complexities became much more difficult to manage.

Review and Study Questions

1. Investigate by yourself how a fender is assembled in a car body. Give special attention to dimensions and tolerances.

2. Investigate all the parameters needed to control a plastic injection process, and explain how each parameter affects the quality of the part.

3. Investigate and propose a plant layout of how a fender production process like MEPO's may be arranged.

4. What are the two technical problems highlighted in this case study? Explain in detail the technical part of the solution.

5. Fill out the following table to clearly show a summary of differences and similarities of French and Mexican cultures.

French-Mexican Culture Summary

Differences		Similarities
France	Mexico	France and Mexico

6. Provide measures to take advantage of culture similarities while working with French and Mexicans in global engineering projects.

7. Provide measures to overcome culture differences while working with French and Mexicans in global engineering projects.

8. Discuss all the implications of line managers using their working time to inspect the molds.

9. What is a Poka-Yoke device? Investigate and propose with a sketch such a device for this case study and give a short operating manual.

10. Investigate what is a greenfield start. What other ways are there to start an international business?

11. What is the difference between engineers being specialists and generalists? Research some other universities' curricula to give a clear differentiation.

12. Investigate and give examples of a company database of solved problems and cases.

Notes

1. A. Rodríguez-Trejo, "La gran aventura del cambio," *Revista Mundo Ejecutivo* (April 2002): 5–7.

2. G. Hofstede, *Culture's Consequences*, 2nd ed. (Thousand Oaks, CA: Sage, 2001), pp. 171–75.
3. F. Trompenaars and C. Hampden-Turner, *Riding the Waves of Culture*, 2nd ed. (New York: McGraw-Hill, 1998); G. Inzerelli and A. Laurent, "The Concept of Organizational Structure" (Working Paper, University of Pennsylvania and INSEAD, 1979) and "Managerial Views of Organizational Structure in France and the USA," *International Studies of Management and Organizations* 13 (1983): 97–118.
4. Hofstede, *Culture's Consequences*; also see Raúl Béjar Navarro, El *mexicano: Aspectos culturales y psico-sociales* (Mexico: Universidad Nacional Autónoma de México, 1987); and Eva S. Kras, *Management in Two Cultures: Bridging the Gap between U.S. and Mexican Managers* (Yarmouth, ME: Intercultural Press, 1989).
5. Hofstede, *Culture's Consequences*, pp. 95–100; A. R. Negandhi and S. B. Prasad, *Comparative Management* (New York: Appleton-Century-Crofts, 1971).
6. J. L. Barsoux and P. Lawence, *Management in France* (London: Cassell Eduational, 1990); P. d'Iribarne, *La logique de l'honneur: Gestion des entreprises et traditions nationales* (Paris: Seuil, 1989); M. Crozier, *The Bureaucratic Phenomenon* (Chicago: University of Chicago Press, 1964); Hofstede, *Culture's Consequences*, especially p. 216. For a "politeness" explanation, see Edward T. Hall and Mildred Reed Hall, *Understanding Cultural Differences* (Boston: Intercultural Press, 1998).
7. To an outside observer it seems obvious that production processes must be adapted to climatic conditions like altitude, temperature, and barometric pressure. However, many technology transfer projects coming to Mexico neglect these factors (E. F. Moritz, ed., *Sports, Culture and Technology: An Introductory Reader* [Bremen: Artefact Verlag, 2003]). In this case study, the barometric pressure, average temperature, and humidity all were significant factors affecting the production of sound plastic parts.
8. Kras, *Management in Two Cultures*.
9. Harry G. Barkema, John H. J. Bell, and Johannes M. Pennings, "Foreign Entry, Cultural Barriers, and Learning," *Strategic Management Journal* 17 (1996): 151–66; B. Kogut and H. Singh, "The Effect of National Culture on the Choice of Entry Mode," *Journal of International Business Studies* 19 (1988): 411–32; J. T. Li and S. Guisinger, "Comparative Business Failures of Foreign-Controlled Firms in the United States," *Journal of International Business Studies* 22 (1991): 209–24.
10. Dianne Vaughan, *The "Challenger" Launch Decision* (Chicago: University of Chicago Press, 1996).
11. Hofstede, *Culture's Consequences*.

5

USAHP Confronts Mexico's Subcultures

USA Home Products (a pseudonym, which we will abbreviate as USAHP) is a highly successful multinational corporation that has been in operation in the United States for more than a century and a half. Today it employs more than 100,000 people in manufacturing plants in eighty countries, which produce home products sold in one hundred and forty countries. Selling more than three hundred product brands, it has a presence in over 75% of the globe and annual revenues of more than US$40 billion. It has operated in Mexico since the end of World War II, with a large and increasing market share for each of its nine divisions. In 1997 it purchased a state-of-the-art production plant from a local company (which we will call MEXCO) in one of Mexico's most distinctive provinces, Tixtlan (also a pseudonym; true name was changed for confidentiality reasons). The plant currently employees five hundred people and produces a million cases of products each month. A key element of USAHP's global vision is maintaining consistent operations across all of its plants, regardless of what they make, who they sell it to, or where they are located. Its management believes that competing in a global marketplace requires a company to have low costs and high quality. Making certain that all of its plants use the same production processes is USAHP's primary strategy for achieving those goals. Consequently, whenever it acquires a plant, distribution center, or office, it quickly moves to install the "USAHP way." This strategy means that employees in an acquired operation immediately experience rapid change, and all of the uncertainties and anxieties that accompany change. Dealing with change, and with resistance to it, is the story of this chapter.

U.S. and Mexican Cultures: So Close, and Yet So Far

As a result of their close geographical relationship and the reduction of legal barriers to economic exchange that accompanied the passage of the North American Free Trade Agreement (NAFTA), Mexico now is the United States' third largest trading partner. (Note: Recently China displaced Mexico from second place.) U.S. investments in Mexico grew significantly during the 1980s and 1990s, and promise to continue to do so in the near future. However, the two countries are very different, not only in terms of obvious factors like wealth, power, and language, but also in terms of national cultures.

Characteristics of U.S. Culture

U.S. culture is characterized by a low uncertainty avoidance index (UAI), moderate to low power distance (PD), and very high individualism (with scores three times as high as Mexico's). Like other relatively wealthy countries with long-term, stable political systems, U.S. workers live in a relatively stable environment. As a result, they feel less stress, are more likely to challenge organizational rules, and are not especially interested in staying with the same company over the long term. They have a high ambition for advancement, welcome competition, value individual, rational decision making, expect hiring and promotion decisions to be based on expertise and credentials rather than personal or political connections, and are most productive when organizations employ reward systems that are based on individual performance. They are less resistant to change, expect to be consulted about decisions that involve them directly, feel comfortable disagreeing with their supervisors, and are more optimistic about the motives guiding companies, at least until very recently. Rules and reward systems should be applied universally to everyone, regardless of who they are or to whom they are related. Communication should be direct, efficient, and clear, rather than ambiguous or guided by unwritten rules of behavior and interpretation.

A Review of Mexican Culture

As Chapter 2 explained, Mexican culture is characterized by the opposite: high uncertainty avoidance, high power distance, and low individualism. Subordinates in high-UAI cultures tend to believe that organizations/institutions are not concerned for the welfare of their employees or members, and tend to perceive organizational actions as potentially harmful to those employees. This level of suspicion can be reduced if an employee has a long-term relationship with an organization that he or she believes is successful and stable, and by a system of organizational rules that are stable over time and consistently enforced. They have a high level of faith in specialized expertise, and obtain a degree of security through interpersonal relationships—families, friends, and coworkers with whom they have had long-term relationships. *Organization-initiated* changes stimulate especially high levels of suspicion and anxiety, especially if they are imposed by people with whom employees do not have long-term relationships, are unstructured, unclear, ambiguous, or make employees' futures with the organization seem to be more tenuous. Even the act of planning for change is negatively valued because it increases uncertainty. Although employees in all cultures resist change, the likelihood of resistance is highest in high-UAI cultures. When planning does take place, it tends to be very detailed and short term in focus, based on a narrow range of relevant information, and performed by specialists. Each of these characteristics reduces the uncertainties involved in the planning process.[1]

Mexico also has one of the highest power distance scores among the nations that Hofstede and other scholars have studied. As important for this case study, the PD scores of relatively uneducated Mexican workers in routine jobs are significantly higher than even the Mexican average. At first glance, this suggests that Mexican workers should readily accept changes imposed from the top of their organizations. However, like UA, PD is more complicated than that. First, in high-power-distance societies, change is acceptable only if it is supported by dominant members of workers' in-groups—families, friends, and long-term coworkers. If it is imposed by members of out-groups, especially foreigners or members of a different race, it is likely to be resisted. Organizational policies and practices must be legitimized in terms of the values of the workers' in-groups, and must be accepted by ranking members of that group if they are ever to be accepted by lower-ranking members.[2]

Finally, Mexico is a collectivist culture in which hiring and promotion decisions are expected to be based on personal, familial, and political connections more than on individual ability and training, workers are more productive and satisfied if rewards are based on in-group achievement, and trust is very high among in-group members. Communication and decision making are expected to be open and honest, especially within referent groups. In sum, there may not be any two more different cultures in the world that share a lengthy border and have extensive political, economic, and social ties.[3]

The Culture of Tixtlan State

Chapter 2 pointed out that even in very homogeneous cultures, there are differences in the extent to which individual citizens act in accord with the attributes of a national culture. For example, the first specialist sent from France to deal with MEPO's altitude problem exhibited a very high level of French individualism and sense of superiority, and relatively little UA. In contrast, the second specialist recognized the importance of being part of an in-group to Mexican employees, and adapted to the related aspects of Mexican culture. In short, the first specialist acted in ways that magnified differences in French and Mexican cultures while obscuring cultural similarities; the second one did just the opposite.

In this case, the relevant variation from national culture did not involve individuals, but instead encompassed an entire subculture. Understanding this concept requires a brief lesson in Mexican history. The residents of Tixtlan state have a proud and unique history, one that involves hundreds of years of conflict with their neighbors. In 1504, the Aztecs attacked Tixtlan in an attempt to force them to become part of their empire. The Tixtlan residents incurred enormous losses of life and property, but resisted so fiercely that the Aztecs were forced to retreat. Tixtlan was one of the few kingdoms in Mexico to never come under Aztec dominance.

When the Spaniards arrived in Tixtlan from Veracruz in 1519, they too encountered fierce resistance. Eventually, Hernán Cortés opted to make a truce with Tixtlan, offering to fight together against the Aztecs in exchange for peace. Tixtlan agreed, but only if it was clear that they would fight side by side with the Spaniards as equals, and that their freedom and autonomy as a nation were respected. Cortez agreed, as long as the Tixtlan residents were willing to recognize the king of Spain (Charles V at the time) as the supreme political authority and the Christian God as the one and only.[4]

With the help of 100,000 Tixtlan warriors, Cortez's forces defeated Cholula and then moved on to the Aztec capital of Tenochtitlan, which surrendered in 1521. Tixtlan residents finally saw its long-time enemy humiliated and embarked on what they believed would be a new way of life. However, by the end of the sixteenth century, the promises made by the Spaniards were all long forgotten, as the natives' political autonomy, economic independence, and social freedoms were gradually lost. In 1786 Spain granted control of Tixtlan to its neighboring province, Azteca (a pseudonym). Tixtlan resisted and was once again granted its sovereignty in 1794, a status that was reinforced in 1857 when its own political constitution was signed. But, Tixtlan residents felt that they had been lied to and manipulated, the basis of centuries of mistrust of foreigners and foreign governments.[5]

Throughout the next two centuries, Tixtlan repeatedly was forced to fight for its sovereignty against its Aztecan neighbors. Over time its political and economic influence waned, so that it eventually was reduced to being a crossroads between the Gulf Coast and Mexico City. Its geographical position put it in the midst of important social and political issues, of peace and war, of progress and crisis, but it was hardly ever a main player. By mid-twentieth century, Tixtlan found itself in a rather weak position. The country's government sought ways to promote the state as a profitable place to invest. A new industrialization period took place and has not stopped ever since, in spite of ups and downs that eventually arise. Today it has a diverse and prosperous industrial environment as well as a steady agricultural base. There are plenty of jobs available, sometimes even more than those that can be fulfilled by locals. Poverty rates are low by Mexican standards, and Tixtlan has one of the highest literacy rates and elementary school graduation rates in the country. The state also has thirty-one higher education institutions that serve more than twenty thousand students and twenty-six technical education institutions that host an additional fifteen thousand students. As a result, there currently is a great deal of movement from other Mexican states into Tixtlan. Historically a cohesive, homogeneous, and distinctive state, Tixtlan now is undergoing more change than at any time in its history. Several popular traditions are being changed or abandoned, some of which have already disappeared.[6]

Consequently, the culture of Tixtlan state is in many ways more Mexican than Mexican culture in general. Historically homogeneous, accustomed to high levels of autonomy, proud of their heritage and their independence, and deeply suspicious of outsiders—both from other countries and from other provinces within Mexico—Tixtlan residents have understandable reasons for feeling uncertain about and resisting change. As a result, a multinational corporation from the United States would need to be especially sensitive to cultural differences if its acquisition of a plant in Tixtlan were to succeed.

The Facts of the Case

USAHP and MEXCO had very different ways of conducting business. As is typical of Mexican firms, MEXCO management believed that its employees should learn by experience and trial and error. They hired local employees, more on the basis of established relationships than on education and expertise. As a result, many of their employees had only an elementary school education. New hires begin to work by performing simple tasks until learning, through observation, to master more complicated ones. This approach is called a "don't ask" process, in which a worker acquires specific skills by imitation, just as he or she learned to speak. Worker skills improve with experience, but not in any systematic way.[7] On the other hand, USAHP relies heavily on systematic training and expertise when it makes hiring decisions, and only hires people who have at least some technical education. It prefers to hire outstanding new graduates of selective universities (mostly private), and it requires them to be very proficient in both written and spoken English. Selection decisions are made through a standard U.S. process: applicants are given a test to measure their ability to solve problems, and then the few who pass that test go through a series of interviews.

As a result, after USAHP took over, getting a job in the Tixtlan plant became much more difficult than before, especially for local people. Tixtlan universities are primarily technical schools and lack the prestige that USAHP wants to see in its new hires. USAHP largely recruited its new administrative personnel from neighboring, more developed, and culturally distinct states, the ones that Tixtlan had fought with for hundreds of years. Workers at the MEXCO plant suddenly were confronted by newcomers who were both outsiders (U.S. citizens or from neighboring states) and different in terms of their backgrounds, education, and experience. Expatriates from the United States were doubly alien—they were U.S. citizens and they also chose to live in neighboring states because of their more attractive climate and educational institutions.

For the managers and engineers remaining from MEXCO, the culture change accompanying USAHP's acquisition was enormous. Most of the employees felt, accurately, that their jobs were in jeopardy because of the remarkable differences in philosophy between the two companies. MEXCO promoted, rewarded, and retained employees based on their loyalty to the firm and to their immediate supervisors, and on interpersonal (usually family) connections. In contrast, USAHP's policy was to hire technical experts to start as supervisors and then promote from within the company up to managerial positions. Similar differences were apparent in the reward systems of the two companies. Under MEXCO, managers' salaries were based on their friendships with the owner or the board of directors of the company. As a result, their salaries were far larger than the USAHP salary structure allowed, given their training, performance, and experience. Meanwhile, most of the technicians (machine operators) had salaries that were far below USAHP's standards. In order to bring the new plant into line with company standards, USAHP froze the salaries of the overpaid managers and fired those whose jobs were primarily based on their personal relationships with the previous owners. The managers whose salaries were frozen were unhappy with the change, but they were happy to keep their jobs. However, these steps raised questions about the fairness of the company's reward system. Employees in all cultures view fairness as an important aspect of worker-organization relationships and react negatively to actions that they perceive are unfair. However, the way in which *fairness* is defined varies across cultures. In Mexico, a reward system is perceived as fair if it reflects status within a group and loyalty to an organization; in the United States, it is fair if it focuses on objective assessments of credentials and measures of individual performance.[8]

Moreover, Tixtlan workers defined rewards in practical terms. Income is a means through which one fulfills one's responsibility to one's family, tradition, and religion. It allows a worker to participate in leisure activities with family and friends.[9] The primary topic of discussion at work is family and leisure. Consequently, making demands on workers' leisure time is likely to be resisted, both because it undermines interpersonal relationships and because it upsets long-established routines that help them manage the uncertainty they face. Even the changes that USAHP management assumed would increase worker satisfaction—shifting from a 6-day to a 5-day workweek (leaving the sixth day for training in new techniques) and building a new cafeteria in which workers could get a complete meal for less than a dollar—were resisted because they upset established routines. Workers continued to buy food from street vendors, often for higher prices, accompanied by the same people they had eaten lunch with for years. When management increased the demands it made of workers (primarily for training in the new production system), they expected to be compensated immediately, rather than at some undefined time in the future. Of course, the company was unwilling to do so because it would increase production costs and undermine USAHP's low-cost, high-quality strategy. Broadening the range of activities required of workers to include administrative duties also heightened the differences in pay between workers and managers, another highly salient aspect of fairness.

In sum, the new owners based rewards, promotions, and retention on objective measures of individual performance. These systems are completely appropriate to the U.S. culture within which USAHP was accustomed to operating. But for the MEXCO employees, the new system created uncertainty and ambiguity and directly threatened their job security. To make matters worse, they were imposed by outsiders, who in many ways had inadvertently exaggerated their alien status. As a result, the changes created an exceptional level of uncertainty for employees whose society is especially focused on uncertainty avoidance.

Implementing the Total Productive Maintenance Concept

The new owners required operators to learn a whole new way of working. They were accustomed to being responsible for a single task—machine operator or mechanic—and were not required to write or document anything. With USAHP they had to change their everyday practices and adhere to strict working standards that required constant documentation. As a result of these changes, many of the older workers accepted voluntary retirement packages because they just could not adapt to the new working systems. The ones who remained found ways to delay or short-circuit implementation of the change.

USAHP uses Total Productive Maintenance (TPM) in its plants to remain competitive. TPM is a manufacturing strategy of Japanese origins that includes different working methodologies. Each company will decide which methodologies to use, but most of them will include the following:

- Focused improvement
- Autonomous maintenance
- Planned maintenance
- Education and training
- Quality-focused maintenance
- Administration and support
- Safety

Autonomous maintenance (AM) is the spinal cord of TPM, and the component that had the greatest effect on the lives of the technicians. The general objective of AM is to hold workers responsible for performing daily maintenance on their machinery. The main objectives of AM are:

- Avoid accelerated deterioration of the equipment by using it correctly and making daily inspections.
- Return equipment to its initial operating conditions.
- Establish the necessary conditions to keep the equipment well maintained at all times.
- Use the equipment as a means to teach new ways of thinking and working.

The AM methodology consists of seven sequential steps:

Step 1. Do initial cleaning.

Step 2. Eliminate all pollution sources and difficult-to-reach places.

Step 3. Establish standards of lubrication, cleaning, and bolt tightening.

Step 4. Inspect the equipment.

Step 5. Inspect the processes.

Step 6. Systematic AM.

Step 7. Self-administered system.

Phase I: The Early Years of AM

In order to implement AM, USAHP started training at the top managerial level. Their goal was to convince the rest of the employees that AM was so important that even managers would get their hands dirty working on the machines. For the start-up of AM, all the leaders of the plant, including the plant manager, cleaned their assigned work centers and documented the status of their work. The next step was to involve all the plant to work in AM. An exchange of personnel from different sites where AM was already in place was promoted. The company provided all the necessary equipment, such as computers, for exclusive AM use. AM was first applied to those operations and equipment identified as most critical for the manufacturing process. Ruben Salas was chosen by the plant leadership team to lead the AM process. He was transferred from another Mexican site where AM had been implemented successfully. Forty teams were formed to pilot the AM project. However, from the beginning Salas had leadership problems with the production department. On the one hand, one would expect top-down change fostered by recognized experts to be readily accepted by members of a high-power-distance society. However, the key personnel involved in this change were outsiders, not members of the operators' existing in-groups.

Arturo Suarez was the production department's manager. He was a local, had worked in the factory for over 10 years, and had gained the respect from the workers in the plant. Suarez started as a supervisor when the plant was owned by MEXCO and worked his way up to his current position. Suarez was under constant management pressure to meet the production requirements set by the sales department. As a result, he opposed having his employees go to AM training sessions because their absence could jeopardize meeting the unit's production quotas. The production workers also noticed that the managers could not come to an agreement on the importance of AM. The new managers had no direct experience on the shop floor, and could not produce persuasive evidence of the positive effects of the system on workers' lives. As a result, the workers decided to listen to Suarez, the boss with whom they had worked for more than a decade and were loyal to. They did not embrace the AM philosophy and thought that cleaning the equipment was pointless because "it would get dirty again anyway."

During this time the company struggled to provide AM training to its employees. The employees were negatively influenced by Suarez, never understanding why they had to go to training. It seemed like a waste of time to them, and they chose not to go or showed up late. In fact, the implementation of AM seemed to penalize the operators because it eliminated their piece-rate reward system, which gave them immediate, tangible rewards for working hard. In addition, they no longer could blame the maintenance group for any production losses due to machine breakdowns. They believed it was unreasonable for the managers to ask them to fix their own equipment when they had no maintenance experience and previously had been served by a specialized maintenance department. Besides, cleaning is a lower-status activity, almost "women's work" in a highly masculine culture such as Mexico's. They also thought there was no reason to change the current methods because "they had worked fine" all this time for MEXCO. In summary, the workers had no incentives to implement AM, and significant disincentives for doing so. As often observed in intact groups with long histories of working together, the resistance was not individual, but collective.[10]

The workers' resistance to AM also was related to the ambiguity of the system itself. There are no predefined detailed recipes on how to apply AM. Instead, the methodology relies heavily on active worker involvement, a common feature of many Japanese management techniques. For instance, the official AM guidelines state that workers should achieve zero defects, but never explicitly say how to achieve that goal. The methodology states that every member of a work group is responsible for finding and implementing new solutions. In contrast, Mexican workers are culturally predisposed to obey established rules and procedures and follow a leader. Both provide stability and protection in uncertain situations. Even when USAHP did try to adapt to the Tixtlan culture, their actions tended to backfire. Because most of the workers from MEXCO had at most an elementary school education, the new managers decided that the work standards and procedures should be created by supervisors and imposed on the workers. However, AM is designed to involve all workers in developing performance standards and procedures. Although planning by specialists is expected in high-power-distance societies, the respect afforded them depends on the expertise they demonstrate, the success of their efforts, and their ability to create detailed designs and short feedback loops that minimize uncertainty. These operators laughed at the standards created by the outside experts because it was obvious that the supervisors did not know how the real world was. Without a demonstration of relevant expertise and success, the advantages granted to superiors in high-power-distance cultures become disadvantages.[11]

During this initial attempt to implement AM, the combination of cultural preferences and a badly designed reward system probably doomed the system from the outset. After 2½ years, only six groups out of forty had progressed enough to be moved to step 2 of the seven-step procedure.

Phase II: Starting Over

Eventually, USAHP's management realized that their efforts to implement AM at the Tixtlan plant had failed. This is important in itself because there is abundant evidence that managers of U.S. firms tend to persist in failing courses of action long after there is abundant evidence that they indeed are failing.[12] Not only was USAHP's management able to admit failure, but they were flexible enough to change the AM implementation strategy in ways that made it more appropriate to Tixtlan culture, yielding better results.

The transformation began when the company implemented a new reward system that reduced the ambiguities in the system. Clearer working standards were created for AM, including a new audit system where everyone was held directly accountable for their participation in AM activities. New procedures and metrics to evaluate the involvement of the personnel in AM were implemented, such as attendance to AM-related meetings and activities. All AM teams now meet each week to share their results and assign each member with specific responsibilities. Each employee's ranking, promotion, and salaries in the company depend on the level of commitment they have shown to AM. The operators still have a lot of pressure to produce, but now they have preestablished hours to receive training in AM.

Rewards also were distributed on a team basis. Teams were rewarded with small prizes for achieving the set goals, and in particular, special recognition is given to those teams completing important milestones in the TPM process. As important, the team itself decides when it is to be audited and what performance criteria should be used in the audit. Before the team gets audited they get a preaudit with the same criteria used by the real auditors. If the team covers at least 90 points out of a 100, then the team is said to be ready for audit. The audit can be done by any plant worker who is already certified in AM, including the plant manager. The new system returned autonomy to the teams, and reduced the uncertainties built into the changes.

In this particular situation workers from MEXCO were tightly bonded by traditional Tixtlan regionalism and Suarez's leadership. Suarez was relocated to a nonproduction area where he could still be accessible in case the teams needed his technical expertise, but could not interact with the workers on a daily basis. Juan Ruiz, an experienced AM manager who is committed to the process, replaced Suarez as the AM team leader and new production manager. Giving the same man both roles helped eliminate the conflicts and uncertainties of the split system. The new system also used existing in-group relationships to the company's advantage. Ruiz's immediate subordinate is Campos, an old MEXCO supervisor who is supportive of the AM initiative, and who trained three technicians who were put in charge of AM in the different production areas of the plant. In turn, these technicians will pass on the knowledge to the workers in their corresponding areas. A new plant leadership team was formed, which included members from all of the production areas. New subcommittees were created in each area composed of a functional leader (a worker with AM experience) and a leader from the work team.

In the incorporation of other TPM components, such as planned maintenance (PM), care was taken to establish a hierarchy of team leaders that had previous experience and were committed to the TPM program. The members of each shift form their AM teams and work on a specific machine in their assigned area. There's a central leader who is one of the three technicians responsible for the areas. Each of the operators is responsible for his or her machine. The responsibility includes operating the equipment, documenting the problems observed, and repairing minor faults. Examples of minor faults are changing belts or plates, tightening bolts, and lubricating bearings. For major defects, the operators prepare a list that is passed on to the central technician. The central technician collects all the lists and sends them to the production department. Whenever there is a general planned stoppage of the plant all AM teams work in the maintenance of their respective equipments. The responsibility of the previous maintenance team was limited to specialized and complicated maintenance activities rather than routine maintenance activities.

Another important change was the consideration of the operators' educational level. In phase I, this became a source of increased uncertainty and resistance. It was critical to the success of the program because TPM requires that workers document processes and process improvement solutions. Once the workers became involved in the program, they began to understand why additional training was important, and realized that they were capable of learning the new techniques. It became possible to apply statistical quality control tools such as control charts for operators to register the total defects found versus the defects repaired in all the plant. Additionally, workers had to fill in a daily form to check if the cleaning standards were being followed. Today there are thirty-eight teams who already have been certified in step 1 and two teams who have been certified in step 2.

Interpretation

Organizational change inevitably generates resistance.[13] In even the best of circumstances, change creates uncertainty and ambiguity, and employees respond to those anxieties in myriad ways, many of which undermine the organization's objectives. Some changes also carry tangible threats, to status, self-esteem, incomes, or job security of particular groups of workers. As a result, even changes that on the surface seem to promise enhanced rewards or working conditions are resisted. When the changes are sudden, significant, imposed from the outside, involve employees from multiple, different cultures and subcultures, or are mishandled by management, the likelihood of resistance increases substantially.[14]

Both USAHP's global mindset and the structures it used to implement that mindset created resistance at the Tixtlan plant (Figure 5.1). Its insistence that all of their plants use the same techniques in order to maximize efficiency drove a series of decisions that overlooked key characteristics of the local history and culture. The structure through which the operation began also increases the likelihood of resistance.

Cultural assumptions influence the way in which multinational corporations approach expansion into a new country or market. There are many possible forms of expansion. As we explained earlier, greenfield starts, in which a multinational sends a small group of expatriates into an area to hire locals and gradually build a business, minimize the likelihood of culture clash. In contrast, foreign acquisitions, in which a multinational purchases a local company or plant, are fraught with potential for culture clash. In the worst-case scenario, the acquiring firm blindly imposes its own way to doing things on members of a very different culture with little consideration of those differences or for the values of employees of the acquired firm:

> Culture clashes are resolved through brute power: Key people are replaced by the corporation's own agents. In other cases, key people do not wait for this to happen and leave on their own account. Acquisitions often lead to a destruction of human capital, which is eventually a destruction of financial capital as well.[15]

FIGURE 5.1
Global engineering model's center ring influence.

As a result, cross-cultural acquisitions tend to fail significantly more often than other forms of expansion, especially when the cultures of the acquiring and acquired firms/plants are substantially different. For a variety of reasons, national culture and mode of intervention are strongly correlated. U.S. firms tend to expand through acquisitions, rather than mergers or greenfield starts. They also tend to place the burden of adaptation on members of the acquired firm/plant rather than accepting that burden themselves or trying to share it equally.[16] In addition, U.S. managers tend to operate on a one-size-fits-all approach to day-by-day operations, insisting on using the same strategies and practices that have proven to be successful in other operations. When workers do resist change, supervisors tend to attribute their opposition to inadequate training or education, even when it is based on insightful analysis of problems with the new program or the way it is being implemented.[17] There are notable examples of U.S. firms successfully overcoming these tendencies, although success often occurs only after an initial period of cross-cultural conflict and communication breakdown. The methods of managing acquisitions most consistent with U.S. cultural norms are likely to create high levels of uncertainty and anxiety for employees whose national culture makes uncertainty avoidance paramount.[18]

Fortunately, there are cultural characteristics that an acquiring company can use to make the transition go more smoothly. For example, in Mexico, establishing a more personal relationship between supervisors and workers is particularly important to establish loyalty between them. Also, the primary motivation for working is to obtain money needed to live a life they can enjoy. Hence, if the company links a new system to an economic benefit for employees, they are likely to work harder to achieve it. Training is fundamental for the implementation of new systems. However, training programs need to be carefully designed. Relative to Americans, Mexicans initially are more reluctant to work with a stranger; however, once trust is developed they will be very loyal to the team. This suggests the importance of training in teams composed of in-group members when dealing with Mexican companies. U.S.–Mexican cultural differences do not doom acquisitions to failure. Instead, they suggest that cultural sensitivity and cultural adaptation on the part of both parties are especially important to the success of the venture.[19]

Interpretation of Phase I

Like the MEPO case study described in Chapter 4, the second ring of our global model was not especially important to the USAHP case study. Supply chains were short and access to resources did not create problems. Market considerations were somewhat important, largely because of USAHP's success in the Latin American market. It has highly efficient plants throughout the area, and a rather large market share for each. As a result, the home products market is highly competitive, which makes it very important that each

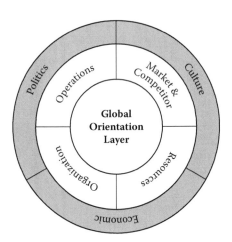

FIGURE 5.2
Global engineering model's outer ring influence.

of the company's plants maximizes quality and minimizes costs. However, the primary complications involved cultural factors (Figure 5.2).

In many ways, phase I of the Tixtlan acquisition provides a textbook case of the problems encountered in cross-cultural engineering. In those cases, a U.S. firm attempts to change operations in ways that conflict with local culture, which inevitably engenders resistance. As Kirkman and Shapiro conclude:

> Change agents, who often are U.S. born because so few host nationals have experience with the latest North American management initiatives (Appelbaum & Batt, 1994), can make cross-cultural mistakes when working in foreign affiliates, leading … to employee (change target) resistance.[20]

However, textbook interpretations of intercultural encounters often overgeneralize cultural tensions and ignore complexities and nuances. In phase I of this case, two complications were especially important.

The first involved the local culture of Tixtlan, which is special even within Mexico. For understandable historical reasons, Tixtlan residents are even more sensitive to uncertainty and more suspicious of change imposed by outsiders than one would expect from a high-UA, high-PD culture like Mexico's. In contrast, the neighboring provinces accepted Aztec control. Today they are an economically wealthier and culturally distinct people who in the past have been enemies of the Tixtlan people. USAHP expatriates chose to live in neighboring provinces for wholly rational reasons—the climate, intellectual attractions, and general quality of life as defined by U.S. culture—but by doing so defined themselves as doubly different. During our study we found no evidence that USAHP's management realized that Tixtlan people are exceptionally sensitive to incursions by outsiders, or that they perceive their neighboring Mexican states and the people who live there as especially alien. Neither did management realize that the efficiency-enhancing steps they had enacted elsewhere upset a tradition of employment and advancement through interpersonal/family relationships that was an important element of Tixtlan culture and the organizational culture of the MEXCO plant. But, when they made those choices and hired new employees from those areas, they exacerbated the problems created by differences between the national cultures of the United States and Mexico.[21]

The second complication involves the concept of power distance. As we explained in Chapter 2, the preference for hierarchy and acceptance of top-down leadership characteristic of high-PD cultures are mediated by the distinction between in-group and out-group ties. Change creates uncertainty. In some cultures the anxiety that stems from change is managed through interpersonal ties. When the changes also circumvent or undermine those ties, uncertainty is increased. USAHP's decision to bring a large number of outsiders, make radical changes in personnel and the hiring/reward system, and suddenly impose a new production system with little explanation magnified the uncertainty further. The initial resistance to implement AM also was related to USAHP's failure to recognize the importance of getting in-group leaders (Suarez) to wholeheartedly accept the TPM philosophy or to recognize that the workers were not responsive to the expertise, status, or power of outsiders. The workers were culturally predisposed to follow Suarez as a local leader.

A third cultural issue disregarded by USAHP in phase I was the orientation that U.S. and Mexican workers have toward working in self-managed teams. In the United States most people leave their parents' home early in their lives, either to go to college or once they get their first jobs after high school graduation. Very differently in Mexico, as in most Latin American cultures, most people live with their families until they get married—even if this is long after they could be financially self-sufficient. So, early in their lives U.S. residents learn to work and trust coworkers with whom they have no familial or extended personal relationship, while Latin American workers are more used to the security of being surrounded and supported by family and *compadres*. As a result, Americans will be more predisposed to rapidly become effective working in teams, while it may take longer for Mexicans to trust their arbitrarily assigned teammates. Conversely, the extreme individualism of U.S. culture makes it difficult for U.S. residents to understand the responses of more communitarian cultures like those of Latin America. In the workplaces of an individualist society others are seen as resources, and tasks prevail over relationships, making teamwork a natural work method. By rapidly shifting to self-managed teams, USAHP maximized uncertainty in a culture in which uncertainty avoidance is highly salient. As a result, it was very difficult to have the workers themselves develop and improve their rules, procedures, and standards as required by the AM methodology.[22]

Interpretation of Phase II

Years of delays in the successful implementation of AM could have been avoided if USAHP's management would have had a better understanding of Tixtlan culture. In particular, the case study illustrates differences among U.S. business and Mexican and Tixtlan cultures with respect to uncertainty avoidance and power distance as they manifest through rules, leadership, teamwork, work ethics, and workforce educational level. The reward system established during phase II did substitute rules for relationships as a source of uncertainty management. Although this constituted a change from traditional Mexican ways of doing things, the workers felt the changes were a welcome improvement over the ambiguities and uncertainties of the past 2½ years. As important, they were implemented in a way that relies on in-group links instead of threatening them.

In the second attempt, USAHP gave more power to the team leaders, and the reward system directly penalized the team members for not participating in the AM program. Kirkman and Shapiro state:

> North American managers are likely to encounter cultures that might be receptive to teams (i.e., low in individualism) but not necessarily to self-management (i.e., high in power distance, being orientation, and determinism) in such countries as Malaysia, Indonesia, the Philippines, and Mexico.[23]

The former type of system, like the one used by USAHP, largely maintains a centralized structure of power and authority. As a result, it is the kind of system that one would expect to be minimally upsetting in a culture with high power distance.

USAHP's management modified their implementation of some AM methods to make them more appropriate to the Tixtlan plant. Since the company's initial goal was to use the same practices in all of its plants, this willingness to adapt is both surprising and a credit to its management. The results suggest that successful implementation would have occurred sooner if those involved recognized the impact of regional cultural differences from the beginning. The Tixtlan plant may never adopt the USAHP way completely. But it is well on its way to implementing a version of it that is appropriate to Tixtlan culture and that fulfills the organization's goals.

Review and Study Questions

1. Explain cultural similarities and differences between the United States and Mexico.

Cultural Differences		Cultural Similarities between United States and Mexico
United States	Mexico	

2. Explain cultural differences between Mexico and Tixtlan.

Cultural Differences	
Mexico	Tixtlan

3. Explain organizational similarities (if any) and differences between USAHP and MEXCO.

Organizational Differences		Organizational Similarities between USAHP and MEXCO
USAHP	MEXCO	

4. Research on your own a more in-depth knowledge of TPM. Also give examples of its applications and results in real companies.

5. Explain in detail AM. This question requires more in-depth research.

6. Define USAHP's global mindset.

7. Research more in-depth all the types of corporate international expansion, i.e., greenfield start.

8. Give some cultural recommendations for American workers coming to work in Mexico.

9. Give some cultural recommendations for Mexican workers hosting Americans in the job place.

10. Define two subcultures belonging to two different regions in your home country.

Notes

1. G. Hofstede, *Culture's Consequences* (Thousand Oaks, CA: Sage, 2001), p. 382.

2. Hofstede, *Culture's Consequences*, pp. 88–90, 97.

3. P. C. Earley, "East Meets West Meets Mideast," *Academy of Management Journal* 36 (1993): 565–81; and "Self or Group? Cultural Effects of Training on Self-Efficacy and Performance," *Administrative Science Quarterly* 39 (1994): 89–117; L. A. Zurcher, A. Meadow, & S. L. Zurcher, "Value Orientation, Role Conflict, and Alienation from Work: A Cross-Cultural Study," *American Sociological Review* 30 (1965): 539–548.

4. R. G. Rendón, *Breve historia de Tlaxcala* (Mexico: Fondo de Cultura Económica, 1996).

5. R. Saldaña Oropesa, *Historia de Tlaxcala* (Mexico: Xicotli, 1950).

6. COPLADET [Comité de Planeación para el Desarrollo del Estado de Tlaxcala], 2000.

7. María Eugenia de la O et al., ed., *Los estudios sobre la cultura obrera en México* (Mexico: Conaculta, 1997); and Rocío Guadarrama Olvera, *Cultura y trabajo en México* (Mexico: UAM, 1988).

8. K. Leung, "Some Determinants of Reactions to Procedural Models for Conflict Resolution," *Journal of Personality and Social Psychology* 53 (1987): 898–908; K. Leung and W. Li, "Psychological Mechanism of Process Control Effects," *Journal of Applied Psychology* 75 (1990): 1134–1140.

9. This also is true of working class employees in the United States. See H. Zoller, "Working Out: Managerialism in Workplace Health Promotion," *Management Communication Quarterly* 17 (2003): 171–205.

10. G. Salancik and J. Pfeffer, "A Social Information-Processing Approach to Job Attitudes and Task Design," *Administrative Science Quarterly* 23 (1978): 224–253.

11. Hofstede, *Culture's Consequences*, p. 382.

12. A. Tegar, *Too Much Invested to Quit* (New York: Pergamon, 1980).

13. S. Deetz, S. Tracy, and J. L. Simpson, *Leading Organizations through Transition* (Thousand Oaks, CA: Sage, 2000); M. S. Poole, A. Van de Ven, K. Dooley, and M. Holmes, *Organizational Change and Innovation Processes* (New York: Oxford University Press, 2000); G. Zaltman, R. Duncan, and J. Holbok, *Innovations and Organizations* (New York: John Wiley, 1973); Y. Zhu, S. May, and L. Rosenfeld, "Information Adequacy and Job Satisfaction during Merger and Acquisition," *Management Communication Quarterly* 18 (2004): 241–70.

14. S. J. Ashford, "Individual Strategies for Coping with Stress during Organizational Transitions," *Journal of Applied Behavioral Science* 24 (1988): 19–36; L. Menzies, *Containing Anxiety in Institutions* (New York: Free

Associations, 1988); M. Mulder, *The Daily Power Game* (Leiden, Netherlands: Martinus Nighoff, 1977); and D. M. Noer, *Healing the Wounds* (San Francisco: Jossey-Bass, 1993).

15. Hofstede, *Culture's Consequences*, p. 45.

16. H. G. Barkema, J. H. J. Bell, and M. Pennings, "Foreign Entry, Cultural Barriers, and Learning," *Strategic Management Journal* 17 (1996): 151–66; B. Kogut and H. Singh, "The Effect of National Culture on the Choice of Entry Mode," *Journal of International Business Studies* 19 (1988): 411–32; J. T. Li and S. Guisinger, "The Globalization of Service Multinationals in the Triad Regions: Japan, Western Europe and North America," *Journal of International Business Studies* 23 (1992): 675–96; Hofstede, *Culture's Consequences*; A. Laurent, "Matrix Organizations and Latin Cultures," Working Paper 78-28 (Brussels: European Institute for Advanced Studies in Management, 1978).

17. K. L. Newman and S. D. Nollen, "Culture and Congruence," *Journal of International Business Studies* 27 (1996): 753–79; T. Zorn, L. Christensen, and G. Cheney, *Do We Really Want Constant Change? Beyond the Bottom Line Series*, vol. 2 (New York: Berrett-Koehler, 1999).

18. J. Mann, *Bejing Jeep: The Short, Unhappy Romance of American Business in China* (New York: Simon & Schuster, 1989).

19. Earley, "East Meets West Meets Mideast."

20. B. L. Kirkman and D. L. Shapiro, "The Impact of Cultural Values on Employee Resistance to Teams," *Academy of Management Review* 22 (1997): 735. Also see E. Applebaum and R. Batt, *The New American Workplace* (Ithaca, NY: ILR Press, 1994).

21. Hofstede, *Culture's Consequences*.

22. Early, "East Meets West Meets Mideast"; Hofstede, *Culture's Consequences*.

23. Kirkman and Shapiro, "Impact of Cultural Values," p. 750.

6

Implementing a Global
Engineering Perspective

The automobile industry was one of the first industries to become a global business. Exporting cars to foreign markets occurred from the very beginning of automobile mass production. Eventually, globalization of production took place. Most automobile firms had multinational operations even before the latest wave of globalization began during the 1980s. This chapter will explain the ways in which product design, development, and production are influenced by factors that are specific to different regions. It concludes with an analysis of a highly successful international operation that has involved two very different societies—Germany and Mexico—for almost a half century. This case study is an exemplar both of how engineers and organizations from very different backgrounds can work together to build a mutually beneficial partnership, and of how difficult it is to build and maintain long-term collaboration in the global economy.

Internationalization of Automobile Production

Automakers have internationalized their operations for a number of reasons. In some cases a company's export operations in large and developed markets such as the United States or Western Europe led the way for moving production abroad.[1] In cases like these, the company wanted to maximize goodwill of the host governments, increase loyalty of their overseas customers, attract additional customers who prefer locally built products, increase their overall production capacity, or avoid trade barriers. This does not mean that they automatically move their engineering divisions overseas; indeed, at the same time that they globalize production they may even centralize product development and corporate functions in their home locations. In other cases, the companies may move operations into emerging markets. Countries like Brazil, China, India, or Vietnam have a very low market penetration (e.g., car density) and a huge population. The growth potential therefore is enormous, which attracts foreign investment. Finally, companies may locate overseas in order to have access to less expensive labor than in their home countries, or be able to operate where taxes and labor or environmental laws are less constraining. For example, European firms moved operations to the southern United States for these reasons, just as U.S. firms moved their operations to Latin America or South Asia. Especially after the development of free trade zones during the 1990s (for example, North American Free Trade Agreement (NAFTA) or the EU), strategic decisions have tended to incorporate many of these considerations simultaneously. Through locating manufacturing operations in countries that are near these large existing markets, such as Mexico, Spain, or the countries of Eastern Europe, companies obtain the advantages of a low-cost environment *and* are able to avoid trade restrictions that exist outside of the zones *and* increase access to new markets.

In addition, the strategies used by companies in one home country may be very different than the ones used by companies in other countries. For example, during the 1980s, mostly Japanese firms had built up production lines in the United States, whereas in the 1990s, prevalent investments had been made by American and Korean firms in the big emerging markets. In this first important phase of globalization, European companies were very conservative with their foreign investments. However, geographical distance increases operational complexity, so firms may decide to simplify the challenges they face by standardizing their products and processes through simplification and outsourcing. Developing worldwide identical—or global— platforms, common processes, and modularization strategies for assembly lines aims toward a global standard. Industry developments of the late 1990s focused on joint international ventures among the strongest companies and global expansion into new markets.

One effect of globalization has been to make it increasingly difficult to identify an automobile as the product of one company or country. General Motors, for example, allied with Suzuki and Isuzu in Japan to sell several products internationally under GM nameplates. In 1998 Daimler-Benz AG merged with Chrysler Corporation. Ford acquired the automobile division of Swedish vehicle maker Volvo in 1999 to be part of Ford's Premier Automotive Group (PAG) with Aston Martin, Jaguar, Land Rover, and Lincoln. Renault and Nissan formed their strategic alliance in 1999, too. A year later GM announced an alliance with Italian carmaker Fiat, which also manufactures cars under the Ferrari, Landa, and Maserati brands. The German carmaker Volkswagen is divided into two brand groups: Audi and Volkswagen. The Audi brand group consists of Audi, SEAT, and Lamborghini, whereas Volkswagen passenger cars, Skoda, Bentley, and Bugatti are part of the Volkswagen brand group. In the years 2006–2007, the main automobile manufacturers, General Motors, Ford, Daimler, Chrysler, and Toyota, together made up to 58% of worldwide production (see Table 6.1). Out of the twelve independent automakers today, nine or ten will likely remain independent during the coming decades.[2]

By the late 1990s, vehicle origins had become so confusing that U.S. consumers, unions, and politicians pressured companies to add statements of domestic parts content to the window stickers on their cars. By 2008, no vehicle sold

TABLE 6.1

World Motor Vehicle Production by Country and Type (2006–2007)

All Vehicles	2006	2007	% Change
Europe	**21,399,289**	**22,845,449**	**+6.8**
• European Union (27 countries)	18,697,868	19,717,643	+5.5
• European Union (15 countries)	16,276,103	16,691,204	+2.6
Double counts Austria/Germany	*−21,501*	*0*	*−100.0*
Double counts Austria/Japan	*0*	*0*	
Double counts Belgium/Germany	*−224,278*	*−196,323*	*−12.5*
Double counts Italy/EU	*−8,214*	*−15,088*	*+83.7*
Double counts Portugal/Japan	*−15,312*	*−18,569*	*+21.3*
Double counts Portugal/Spain	*−27,806*	*0*	*−100.0*
Austria	274,907	228,066	−17.0
Belgium	918,056	834,403	−9.1
Finland	32,746	24,303	−25.8
France	3,169,219	3,015,854	−4.8
Germany[a]	5,819,614	6,213,460	+6.8
Italy	1,211,594	1,284,312	+6.0
Netherlands	159,454	138,568	−13.1
Portugal	227,325	176,242	−22.5
Spain	2,777,435	2,889,703	+4.0
Sweden[b]	333,072	366,020	+9.9
United Kingdom	1,649,792	1,750,253	+6.1
• European Union (new members)	2,421,765	3,026,439	+25.0
Double counts Slovakia/Czech Republic			
Double counts Slovakia/Germany	*0*	*0*	
Czech Republic	854,817	938,527	+9.8
Hungary	190,233	292,027	+53.5
Poland	714,600	784,700	+9.8
Romania	213,597	241,712	+13.2
Slovakia	295,391	571,071	+93.3
Slovenia	153,127	198,402	+29.6
• Other Europe	1,713,641	2,028,392	+18.4
Serbia	11,182*	9,903*	−11.4
CIS	1,702,459	2,018,489	+18.6
Double counts Ukraine/world	*−224,492**	*−257,754**	*+14.8*
Russia	1,503,469	1,660,120	+10.4
Belarus	25,279*	27,708*	+9.6
Ukraine	288,203	402,591	+39.7
Uzbekistan	110,000	184,900	+68.1
Turkey	987,780	1,099,414	+11.3

Continued

TABLE 6.1 (*Continued*)

World Motor Vehicle Production by Country and Type (2006–2007)

All Vehicles	2006	2007	% Change
America	**19,064,649**	**19,109,213**	**+0.2**
• NAFTA	15,909,007	15,454,212	−2.9
Canada	2,571,366	2,578,238	+0.3
Mexico	2,045,518	2,095,245	+2.4
United States	11,292,123	10,780,729	−4.5
• South America	3,155,642	3,655,001	+15.8
Double counts Venezuela/world	*−140,700**	*−143,691**	*+2.1*
Argentina	432,101	544,647	+26.0
Brazil	2,611,034	2,970,818	+13.8
Chile	5,685	10,804	+90.0
Colombia	50,659	73,667*	+45.4
Ecuador	25,170	26,338	+4.6
Peru	0	0	
Uruguay	0	0	
Venezuela	171,693	172,418	+0.4
Asia-Oceania	**28,189,508**	**30,655,981**	**+8.7**
Double counts Asia/world	*−100,000*	*−110,400*	*+10.4*
Double counts China/world	*−80,000*	*−105,000*	*+31.3*
Double counts Thailand/world	*0*	*0*	
Australia	331,768	334,617	+0.9
China	7,277,899	8,882,456	+22.0
India	2,016,511	2,306,768	+14.4
Indonesia	297,062*	412,788*	+39.0
Iran	904,500	997,240	+10.3
Japan	11,484,233	11,596,327	+1.0
Malaysia	502,973	441,661	−12.2
Pakistan	157,514	169,861	+7.8
Philippines	41,603	49,492	+19.0
South Korea	3,840,102	4,086,308	+6.4
Taiwan	303,229	283,039	−6.7
Thailand	1,193,903	1,287,346	+7.8
Vietnam	18,211	23,478	+28.9
Africa	**569,529**	**542,053**	**−4.8**
Double counts Egypt/world	*−30,377**	*−35,200**	*+15.9*
Double counts South Africa/world	*−115,123**	*−104,919**	*−8.9*
Botswana	397	0	−100.0
Egypt	91,518	103,552	+13.1
Kenya	615	705	+14.6
Libya	0	0	
Morocco	28,620	36,671	+28.1
Nigeria	3,000*	3,072*	+2.4

TABLE 6.1 (*Continued*)

World Motor Vehicle Production by Country and Type (2006–2007)

All Vehicles	2006	2007	% Change
Sudan	0	0	
South Africa	587,719	534,490	–9.1
Tunisia	1,740	2,071	+19.0
Zimbabwe	1,420	1,611	+13.5
Others	0	0	
Total	69,222,975	73,152,696	+5.7

Source: Organisation Internationale des Constructeurs d'Automobiles (OICA) (www.diverwyman.com/category/production-statistics).

* Estimate.

[a] Official figures include Belgian GM assembly.

[b] Official figures take account of Swedish manufacturers' world production; in this report, we only use the vehicles produced in Sweden, and the vehicles for which Volvo Trucks does not specify the country of production.

in the United States had domestic parts content greater than 85%, and the figure for most was declining. Some declines were modest—from 90% to 85% for the 2007 and 2008 Chevrolet Silverado pickup truck, or 65% to 50% for the 2008 and 2009 Toyota Corolla, for example—while others were sizable. The Ford Escape fell 25 percentage points (from 90 to 65) when it was redesigned for 2008. In rare cases, domestic content increased, as with the Honda Odyssey (70% to 75%) and Honda Civic (55% to 70%), but the trend clearly is toward reduced domestic content, especially among U.S. manufacturers.[3]

As shown in Figure 6.1, the industry is becoming increasingly consolidated, in terms of both the number of suppliers and the number of automakers. With rising cost pressures, but the requirement of new competencies,

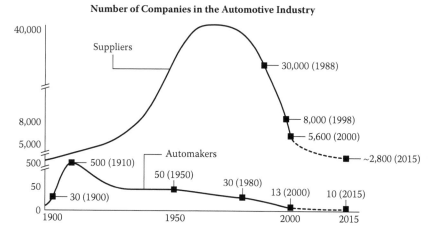

Number of Companies in the Automotive Industry

FIGURE 6.1

Concentration process in automobile industry. (Source: Dannenberg and Kleinhaus, see note 2).

and the resulting need for investment and hence capital, larger and fewer firms are favored. Consolidation increases efficiencies, but it also makes the global system more vulnerable. By the time of the 2008 industry crisis in the United States, suppliers were owed $13 billion by the big three U.S. automakers. Eighty percent of the suppliers that Ford used also had contracts with GM and Chrysler, and one-quarter of its highest-volume dealers also owned GM or Chrysler dealerships. Consequently, a bankruptcy of one company threatened the entire industry, both in the United States and worldwide because of global interconnectedness.[4] The number of suppliers is predicted to be cut in half by 2015, further increasing efficiency and vulnerability.

Localization Influence Factors

In Chapter 1 we introduced the concept of localization and indicated that there are many factors that encourage an industry to localize its production and products. Some of them involve natural features—climate, topography, access, and quality of resources. Others are human creations—the nature of the economic system, infrastructure, level of income and wealth and their distribution throughout a population, demographics such as family size and travel patterns, and the political system, including labor and environmental regulations. Still others involve local culture—language, religion, and factors such as power distance or individuality-collectivism. Localized products better fulfill the specific market demands, and therefore are more competitive. "When introducing products in international markets, weighing the benefits of standardizing products across country markets versus adapting them to the differences among markets is often a significant concern to multinational companies."[5] To be successful, the localized product must lead to sufficient incremental revenues, e.g., through increased sales due to their higher competitiveness, to compensate the higher costs for developing, manufacturing, and marketing that result from adaptations.[6] Arrayed against these pressures to localize are the tangible advantages of standardization and economies of scale. For automobile manufacturers there are many reasons not to localize their products. Higher developing and manufacturing costs for product adaptations can be avoided by standardized products. Each organization must find a way to balance these pressures in each of the situations it faces.

Localization of Products

Localization causes extra costs for manufacturers, especially when exporting products to foreign countries. Homologation aims at following local laws and obtaining necessary certificates at the lowest costs possible. Localization, however, is far more complex. The first step in adopting optimal localization

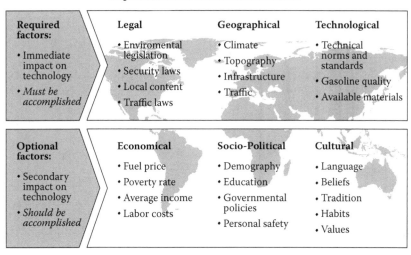

Important Localization Influence Factors

Required factors:	Legal	Geographical	Technological
• Immediate impact on technology • *Must be accomplished*	• Enviromental legislation • Security laws • Local content • Traffic laws	• Climate • Topography • Infrastructure • Traffic	• Technical norms and standards • Gasoline quality • Available materials

Optional factors:	Economical	Socio-Political	Cultural
• Secondary impact on technology • *Should be accomplished*	• Fuel price • Poverty rate • Average income • Labor costs	• Demography • Education • Governmental policies • Personal safety	• Language • Beliefs • Tradition • Habits • Values

FIGURE 6.2
Required and optional localization influence factors.

strategies is distinguishing between required and optional localization factors (Figure 6.2).

Required factors, which make it necessary to localize a product to a specific foreign market, include geographical factors and technological norms and standards. Differences in voltage or in the metric system are common adaptation needs. The most important and obvious need for product localization, of course, is different government regulations in foreign markets. Governmental regulations and norms often increase with time, like emission regulations, resistance of bumpers, or the type of air bags for automobiles. Optional localization seems more interesting, although it is more complicated, because the parameters are not clearly available, like in the case of required localization, where technical requirements of governmental regulations are well defined. Income levels, different customer preferences, different cultural value systems, and the level of education and technical abilities may differ from country to country.

When evaluating the advantages of localization, the differences between required and optional localization have to be considered. When there is no other way than to adapt the product to the local requirements, for example, to fulfill local legislation or to ensure basic functions, the advantage of product localization is simply the possibility to enter the market. Voluntary adaptation aims at a better-adapted product, which is able to increase market acceptance and customer satisfaction. Localized products will better fulfill the customers' preferences and will add additional value to the product because of its increased usability in foreign circumstances. A localized product will have a competitive advantage in the market, and thus generate higher margins or more sales and revenues than a nonadapted product. In today's automobile industry, fulfilling the environmental and safety legislation is mandatory. However, as cars normally are designed for the United States, Europe, or Japan, they all from the beginning do fulfill high standards. In the case of developing countries like Mexico, where environmental and safety legislation are not as demanding, nonlocalized cars would often overfulfill these regulations. In this case, high-elaborated technology can be taken out of the car or substituted with more economic solutions. The resulting cost savings can be used to improve the market position of the car through lowering its retail price or to realize higher margins.

Even with flexible manufacturing systems, rapid prototyping, and just-in-time and just-in-sequence methods, the costs of offering localized products in various markets is higher than offering a global product to multiple markets. In development, additional effort is necessary to localize the product. In manufacturing, there are disadvantages if it is not possible to use the same raw materials, same machines, same tools, and same processes. In both development and production, the cost of increasing variety (e.g., complexity costs) due to localization has to be considered. Standardized products can lower costs through economies of scale, which can be lost due to localization.

The balance between localization and standardization is also influenced by the product itself. For example, the balance is different between premium and economy cars. Upmarket car producers make relatively few changes to their products, whereas economy cars have to be localized in more aspects to be successful in foreign markets. Premium cars have to be localized to fulfill requirements, technical standards have to be fully accomplished, and geographical circumstances have to be considered. Safety standards such as lightning regulations have to be fulfilled completely. Because premium cars are built to compete at least in one of the most demanding markets, such as the United States, Europe, or Japan, they normally do overfulfill legal requirements when exported to developing countries. Technical and geographical circumstances, like fuel quality or climate, have to be considered as well in premium cars. For example, their high-developed state-of-the-art engines often have to be modified or changed for an older model to cope with lower fuel qualities. Bad road conditions afford underbody protection, regardless of the type of car. Table 6.2 shows a comparison of the localization of premium and economy cars.

Culture distinguishes between economy and premium cars. Premium car brands already reached a global identification with their products due to their strength and international success through the image, quality, and performance of their products. Moreover, car users would be upset if global luxury car brands change their cars. For example, customers want to buy a German sports car, and therefore do not accept differences, even if they are more appropriate to local conditions. Luxury has become global. Only if, for example, a BMW of the 7-series is marketed in a country where the owner of this prestige product is usually being driven and sitting in the back do far-

TABLE 6.2

Localization of Premium and Economy Cars

Localization	Premium	Economy
Legal	All requirements have to be fulfilled.	All requirements have to be fulfilled.
	Overfulfillment of requirements is acceptable and often expected by the client.	Through preventing overfulfillment of legislation, a lower retail price can be realized.
Geographical	Premium cars have to be adapted because of sensible technology.	Economic cars have to be adapted to external circumstances (road conditions).
	Premium cars have to be adapted to external circumstances (road conditions).	
Technological	Low production volumes.	High production volumes.
	Global suppliers are used.	Local suppliers are used.
	Electronic equipment has to be localized.	
Economic	Retail price is not a critical purchase criterion.	Income levels are very important.
		Adaptation of the car to ensure competitive retail price is crucial for market success.
Sociopolitical	None.	None.
Cultural	Global brand image overrides local customer preferences.	Local culture is important.
	Luxury is global.	Adaptation can generate additional value for the customer.

reaching changes in interior decoration and equipment become necessary. Of course, the operation of media, air conditioning and heating, windows, etc., primarily has to be organized from the back.

Localization in the Product Development Process

Products are normally developed with a strong focus on their target market. A systematic approach to integrate localization in the product development process is not used in the automobile industry. Product development process in the automotive industry is shown in Figure 6.3. In fact, localization is done in various stages of the product development process, by different groups with different goals. In the automobile industry, product development is highly time critical and very complex. A car is normally designed for use in a specific country or region. However, all future export destinations and their regulations have to be considered from the beginning. In car development, a car equipped with all imaginable extras is always the base model for all calculations.

FIGURE 6.3
Product development process. PD, project definition; PA, project authorization; SDP, series development and preparation; SOP, start of production; ME, market entry.

Localization is done just before the start of production (SOP), where adaptations necessary for a specific market entry (ME) are made. Normally, the best-equipped calculation model will be adapted to the local market in taking out all technology that is considered not useful. Other possibilities for localization are model changes and product cost optimizations. A car model is changed a few times in its life cycle to remain attractive. The exterior design and the basic equipment package will be changed. It is also an opportunity to localize the car. Product cost optimization is responsible to reduce the cost of the product throughout its life cycle. In this optimization cycle, country-specific solutions are also considered.

An alternative way of product development is proposed by Rauner.[7] In this approach a product is developed that is as culturally and technically neutral as possible. This phase is called internationalization—the opposite of the localization that comes later in the process. This multistage process reduces the time and resources required for localization, saving producers' money and improving their time-to-market abroad. Culturally neutral products (CNPs) are not ready-for-use goods. They are basic technologies or basic products. The core technology is kept neutral, so in order to apply a product for a certain market, additional region-specific technology or adaptations have to be made. This is a considerable difference to the common development process in the automobile industry. It will be subject to discussion on whether a culturally centered approach could lead to a less complex process, able to cope better with the goal of localized products.

Localization in Its Most Extreme Form: Customization

The most extreme form of localization is mass customization or build to order. Like localization, customization requires a high degree of flexibility in production processes. Build-to-order is defined as building a car according to the customer's specifications while there is a definitive order in place. Demanding customers, e.g., with their desire for more individuality, and the hard competition in the automobile industry have led manufacturers to consider implementing build-to-order models. Build-to-order is not a new concept. In the United States, about 7% of cars ordered are custom-made. In Europe, about 19%, and in Germany even, about 60% (even if not always customized) are build-to-order cars.[8] To achieve benefits of mass production and in order to offer individual specification possibilities without giving up the

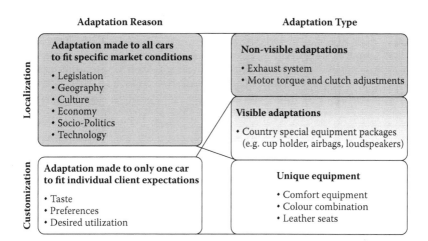

FIGURE 6.4
Differentiation of localization and customization.

goal of a rapid delivery time, it is necessary to leave customized pieces to the end of the assembly process, for example, standard engines whose performance characteristics can be determined in the final assembly line through changes of the engine control module's software.

Mass customization is defined as "producing goods and services to meet individual customer's needs with near mass production efficiency."[9] It makes use of flexible computer-aided manufacturing systems to produce custom output, reaching low unit costs of mass production processes with the flexibility of individual customization. It seems logical that high-end premium cars are provided with high levels of customizations, whereas lower-end economic cars are only given low levels of customization. While hundreds of configuration options may make sense for premium cars, they rarely do for economy or compact cars. Manufacturers attempt to deal with the trade-off between flexibility (i.e., many variants) and efficiency (standardization) by using modular assembly processes and easily interchangeable parts, even if it means putting a BMW engine in a Peugeot body. Customization and build-to-order are dominating strategic discussions about the future of the automobile industry. But even implemented completely, there will be a need for localization. Customers like to have the possibility to choose their engine, exterior and interior colors, and comfort and safety equipment. In the future, they might want to choose even more technical details, e.g., the supplier of the car's suspension. Even customized parameters of programmable features of a car seem possible. Differences between localization and customization are shown in Figure 6.4.

Platform and Module Strategies

A platform is comprised of the low structure of a car. The main objective of this structure is to provide support to the rest of the vehicle. Almost all the components of a platform are metallic parts welded or fastened to the main supporting structure. A platform is rarely seen by the consumer, so variety for customization reasons is seldom necessary. However, platform variety arises from the need to accommodate a different car body, engine size, transmission, or suspension. Using the platform strategy, design and manufacturing of vehicles is enhanced since a number of different car models can be produced to satisfy several market segments and maintaining an efficient production scale. For instance, in 2002 Nissan reduced from twenty-four to five platforms to produce almost all its vehicles, with a savings of 40% in the number of assembled parts. From the design standpoint, very costly engineering work is recycled so that vehicles with low production volumes can still be manufactured.

A module is defined as a component assembly, representing a functional and logical entity, which—for that reason—can be replaced completely as a whole. Modularization is understood as the process to combine component assemblies, system components, and individual parts into one module. For example, typical modules in a car are the front car, driver's cockpit, body, indoors, and car floor (Figure 6.5). Strategic goals are the enhancement of product quality through one quality-responsible module (or system) supplier, higher efficiency through simplification of assembly operations, higher functionality through integration of single parts and functions, and higher productivity through reduction of development and coordination of time through shifting system responsibility to the module (system) supplier. On the other hand, the carmakers lose control over their processes, and risk

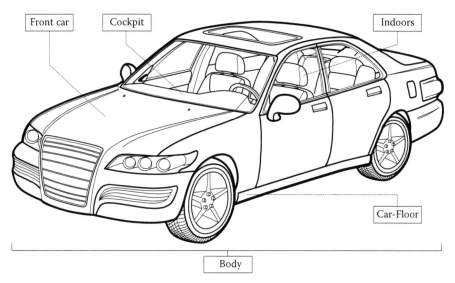

FIGURE 6.5
Modularization of a vehicle.

being converted into simple assemblers of technologies developed by their tier 1 suppliers.

Upcoming Strategic Challenges

Although our summary so far may make it seem that the automobile industry faces a bewildering set of challenges at present, it will have to undergo further consolidation and cope with additional shifts in pressures as globalization continues to develop. "The winners will be those companies that build up new competencies: software development, mechatronics or digital supply chains, as well as new social and cultural competencies within the framework of globalization."[10] Alongside the need to continue to lower costs and cope with competition, automobile companies see increasing customer requirements regarding quality and safety of their cars, especially in highly demanding markets such as Germany or Japan. Furthermore, development of cars is increasingly guided by the more stringent environmental standards in many countries. "In the next 10 years, cars will become about 30% quieter, fleet consumption will fall by 15% and, thanks to new engines and catalytic converters, the output of noxious substances will amount to only 1/1000 of what was considered state-of-the-art just three vehicle generations ago."[11] A recent study has revealed the upcoming technological innovations, which will have an impact on the automobile industry. An overview is given in Table 6.3.

The response of automobile manufacturers to the growing diversity of models and versions is to devise new solutions in the field of vehicle construction. The segmentation of the car into four modules (passenger compartment, front, roof, and rear modules).

However, the main challenge will be technological innovation, from microelectronics to innovative materials to production technologies. The fuel cell is coming, but not until 2015. Until then, the new technology will be tested in mini-production series. Overall, however, the proportion of vehicles powered by alternative drive concepts (gas, electric, fuel cell) by 2010 will be only 10%. Reducing a vehicle's weight by 100 kg will reduce fuel consumption by approximately 0.8 L per 100 km. Such efficiency calls for the innovative use of materials such as high-strength steels, metal foams, magnesium, ceramics, and aluminum. By 2010, weight will decrease by 17%, or by an average of 250 kg per vehicle. Engineers, especially in Germany and Japan, will drive much of this change. They already have a decisive influence on the success of their respective domestic automobile industries. Ongoing competition between production technologies, the integration of components and functions, optimization across the entire production process, and continual improvements in precision will enable the national automobile industries in these two countries to remain one step ahead of the rest of the world.

TABLE 6.3

Technological Innovations in Automobile Construction

Year	Chassis	Power Train	Engine and Auxiliary System	Body Structure	Body/Exterior	Interior	Electrics/Electronics
2001	Ceramic brakes Run-flat technology SWT sensory	AMT-CVT Magnesium gearbox	Ceramic glow plug Fully variable mech. valve timing Diesel high-pressure supercharging Gasoline direct injection	Metal foams Steel space frame Sandwich structure	Aluminum Plastics Magnesium Hydrophobic surface Active lighting	Soft-touch surfaces Smart airbags LED technology	Opitcal bus systems Hybrid vehicle electrical system
2005	Active chassis Magnesium wheel suspension Electromechanical brakes Steer-by-wire Plastic wheel suspension	Starter-generator Infinitely variable transmission Double-clutch gearbox	Particulate filter Intermetallic materials Electric coolers and air control/cooling Urea catalytic converter Denox storage cat. conv. Electric hybrid drive Fuel cell drive Electromechanical valve gear Electrohydraulic valve gear	Composites Plastic body	Surrounding area detection with radar Cameras for object detection Pedestrian protection sensors	Interior lighting with central light source Night vision Fully variable interiors	42V vehicle electrical system Standard operating system Pre-crash sensors Heads-up display Car PC
2015		Wheel hub drive	Hydrogen combustion				Driving with auto pilot

Source: Mercer Management Consulting, *Automobile Technology 2010* (www.oliverwyman.com/ow/insights).

A Case Study in Global Automotive Engineering: The German-Mexican Partnership

Although General Motors and Volkswagen opened their first Mexican plants during the 1960s, a decade later Mexico's motor vehicle industry still was characterized by outdated machinery making products that could not compete successfully in the international market. Since 1990, foreign automakers, attracted by wages as low as $1.50 an hour, have invested an average of $2 billion per year in Mexican operations.[12] By the end of the 1990s, the technology in Mexican plants allowed them to compete success-fully in the worldwide market.[13] Today, foreign auto manufacturing plants (see Figure 6.6) employ a half-million people, directly or indirectly, and account for a fifth of the country's exports. Mexico has become a global purchaser and supplier of passenger cars and commercial vehicles. With annual motor vehicle production close to 2 million units, it is the eleventh biggest car producer worldwide and the region's second biggest producer after Brazil. A wide variety of vehicles are produced, in plants located throughout the country.

The big four in the Mexican market are General Motors, Volkswagen, Nissan, and Ford, which together account for over 75% of the national sales

FIGURE 6.6
Locations of the Mexican automobile industry.

(U.S. companies account for more than half of the total). Successful cars are the General Motors Chevy, the Nissan Tsuru, and the Volkswagen Pointer, all subcompact economy cars. Three-quarters of the vehicles made in Mexico are exported. Traditionally, almost all of Mexico's auto exports went to the United States and Canada (96% in 2006), but that figure has fallen significantly (to 77% in 2008) as manufacturers have sought to diversify the Mexican export business (see note 12). Currently, only about 19% of the Mexican population can afford to own a car, but the Mexican market is still expanding, with sales exceeding 1 million units by the mid-1990s.

In contrast, Germany is the world's fifth largest economy and the third biggest car producer after the United States and Japan. The German industry is highly diverse. Germany is home to some of the world's premium car manufacturers—BMW, Mercedes Benz, and Porsche—and is seen as one of the most demanding car markets worldwide. But, its biggest carmaker, and Europe's biggest carmaker, is Volkswagen. General Motors, Ford, and Toyota also have production facilities in Germany. The BMW Group, with its brands BMW, Mini, and Rolls Royce, is headquartered in Munich. The only plant outside Bavaria is located in Berlin, where motorcycles and vehicle parts are manufactured. The Daimler Group is the world's fifth largest car producer in terms of sales.[14] Its European headquarters is located in Stuttgart. The passenger car brands include Maybach, Mercedes Benz, and Smart. Commercial vehicles, including light- and heavy-duty trucks and city and overland buses, as well as engines and parts are fabricated in Germany. The Volkswagen group, headquartered in Wolfsburg, comprises Volkswagen passenger cars—such as Beetle, Golf, Polo, and Lupo—Skoda, Bentley, and Bugatti, while the sport Y Audi brand group includes the Audi, Seat, and Lamborghini brands. Volkswagen-Nutzfahrzeuge sells various commercial vehicles under its name, such as vans and light trucks, buses, and pickups.[15] General Motors' Buick, Cadillac, Chevrolet, Pontiac, Saab, and Vauxhall brands also have a strong presence, as does GM's German subsidiary, Adam Opel AG. The Ford Motor Company produces the Ford Fiesta at its German headquarters in Köln and the Ford Focus in its Saarlouis plant. The plants of the exclusive carmaker Porsche are located in Leipzig (Cayenne, Carrera GT) and Stutgart-Zuffenhausen (all other models).

An engineer's training and experience usually are culture specific. He or she knows the conditions at home, so much so that it becomes taken for granted. In much the same way that a fish (probably) does not realize it is in water, engineers often forget that the circumstances surrounding them are distinctive—other countries have different conditions, societies, political systems, economic realities, and so on. And, engineers from those countries also take their surroundings for granted. As Chapter 2 explained, the first step in becoming a global engineer is becoming conscious of one's own surroundings, and the fact that they may be very different than those experienced by others.

Localization Influence Factors and the Mexican Auto Industry

In order to localize a car to foreign environments, many factors have to be considered in the design and production of automobiles. As we explained earlier in this chapter, there are a number of factors that corporations must consider when they plan to operate in or sell products to foreign countries. Some of these required factors result from natural conditions—geography, topography, and climate. Others are created by human beings—the transportation infrastructure and traffic patterns. Of course, natural conditions interact with human social and cultural development in complicated ways, but they also have a direct effect on product design and production processes.

Required Factor 1: Geography

Mexico covers a territorial area of 1,964,375 km², and it is the fourteenth biggest country in the world and has an amazing variety of topography and climate, all of which place demands on automobile design and production. There are coastal plains and the large high plateau of central Mexico, which is surrounded by two major mountain chains: the Sierra Madre Occidental and the Sierra Madre Oriental. The central plateau historically has been the home of most Mexicans. Mexico City, the capital and its suburbs, constitutes a huge metropolitan area that dominates the rest of the country's economic, political, and cultural life. One-third of the population and 80% of the Mexican automobile market are concentrated in this central high plateau. Mexico also has a more varied climate, from deserts in the north to tropical rain forests in the south, from the hot and humid conditions at the coastal plains to the drier and mild conditions at the central high plateau. The temperatures on the high plateau are mild. The average temperature is about 23°C. The hot and rainy season begins in May and ends in October in most parts of the country. November to April is the dry season, whereby December and January are the coldest months, with possible morning temperatures lower than 0°C on the high plateau. Mexico City has an average of 2,428 sunshine hours per year, with an annual global radiation of 6,642 MJ/m².

In contrast, the Federal Republic of Germany is one-fifth as large as Mexico (357,021 km²) and has a less diverse topography. More urbanized than Mexico, it is bordered to the north by the Baltic Sea, the North Sea, and Denmark; to the south, Austria and Switzerland; to the east, Poland and the Czech Republic; and to its west, the Netherlands, Belgium, Luxembourg, and France. The capital, with 3.4 million inhabitants, is Berlin.

Climatic factors have an important effect on automobile design in the two countries. Neither country is especially hot or cold, reducing the need for localized designs of engine cooling systems or heating and air conditioning systems. Examples of hot countries are many African countries (e.g., South Africa) and Australian regions; whereas Russia (particularly Siberia), the northern Scandinavian countries, and Alaska can be considered cold areas. Design measures for hot countries include changes in the cooling system to provide higher cooling performance. Therefore, changes in the electric power supply could become necessary. Often in cold countries, the steering gear oil has to be changed to provide appropriate viscosity. Measures against snow blockage have to be taken. Furthermore, different from Mexico, in Germany the coolant and washer fluid have to be frost-proof down to −30°C. To cope with the frequent snowfall in Germany, cars are also equipped with all-season tires or even with two sets of tires, one for summer and one for winter use. The correct functioning of the air conditioning system in countries with high atmospheric humidity has to be ensured through a modification of the system response curve. In the same way, appropriate heating power has to be ensured.

Because temperatures in Mexico normally do not get below freezing, the cold start criterion, that is, the minimum temperature at which the engine has to be able to start, is only −10°C. In Germany, a cold start has to be possible at temperatures as low as −30°C. To start an engine, a very powerful electric starter motor is required. The energy necessary is delivered through a starter solenoid by the car's battery. This starting system could be adapted to the easier requirements of Mexico in order to save costs. Plastic parts and cables are to be dimensioned for the temperature range as well. Extreme cold as well as extreme heat can make these parts brittle and fragile. For example, the wiring harness in the engine hood due to high temperatures and high global radiation has to sustain great heat in Mexico. Rubber seals of doors, interior plastic parts, and the painting of the car have to withstand this radiation as well.

Both countries have varied topographies. Mexico's high plateau covers almost all of the country and has an average elevation range between 1,500 and 2,500 m, with the volcano Pico de Orizaba (5,700 m) as its highest elevation. With many cities situated in the high mountains, the inclination of the streets can reach high values. Germany's territory consists of lowlands in the north, the forested uplands in the center, and the Bavarian Alps with the Zugspitze (2,963 m) as its highest elevation in the south. Nearly all of Germany's traffic takes place at an altitude of less than 500 m above sea level, and the elevation almost never reaches more than 1,500 m.

Due to the altitude of the Mexican highlands, most of the cars are used at a low static air pressure. Due to the lower air density, the volumetric airflow is limited. Due to the given stoichometric mixture between air and fuel, less air intake leads unavoidably to less air–fuel mixture. With less fuel intake, the engine's output will be lower. Therefore, the available engine torque and the performance of a car in Mexico City (2,300 m) is about 20 to 30% less than it is near sea level. While old cars with carburetors need adaptation at high altitudes, modern fuel injection cars, detecting the air–fuel ratio with lambda sensors and calculating the fuel needed in their engine control unit, will easily balance the low air density. Because of the resulting lower performance of the engine, the engine's workload will be even lower than normal. To reach a good drive ability of the car, it is appropriate to change the car's rear gear ratio by using an adapted differential to ensure a sufficient acceleration. However, the expectations of clients demanding huge acceleration performance are not fulfilled this way. Therefore, the use of super- or turbochargers, which through compressing intake air are not dependent on ambient air density, has to be considered. Turbochargers are very common in Mexico. Similarly, cars without turbo charging in Mexico often use a bigger engine than their German counterparts.

One characteristic of Mexican streets is a high inclination. Together with the omnipresent speed bumps, they force drivers to often slow down and accelerate on streets with very high inclinations. This situation, together with the lower engine torque due to the altitude, is the reason to design the clutch with a different maximum workload.

It is not surprising that the infrastructures of Germany and Mexico are quite different. Mexico has a good—although rather expensive—toll express freeway system throughout the country with a length of 6,429 km. The freeways are usually in good condition. The whole road system has a length of 329,532 km, of which only a third (108,087 km) is paved. There are many streets in bad condition. Furthermore, one can find many speed bumps (locally called topes), which constitute a serious danger for the car underbody, exhaust system, and axle. Basically, anyone can put in a tope of any type on any road to control the speed of the traffic. This includes everything from U.S.-style speed bumps to mounds of dirt piled up on the highway to large fishing ropes. Most, but not all, topes are marked, and some topes that are marked do not actually exist. Many holes in the asphalt surface or damaged manhole covers create dangers as well. Complete underbody protection or at least protections for the oil pan and gearbox have to be applied. Furthermore, reinforcement of shock absorbers has to be considered. For nearly all passenger cars, a lift-up of the chassis is necessary to gain greater ground clearance. "The critical clearance height for the poor road conditions is 170 mm."[16] For instance, Volkswagen de México increases the standard clearance to 10 mm for cars sold in Mexico. On Mexican highways, a speed of 80 to 120 km/h is adequate. On express freeways, a speed of 120 to 140 km/h is usual. Cars in Mexico therefore do not need aerodynamic parts and high-performance brakes. These parts may, on the contrary, be very vulnerable, due to the bad street conditions. If not considered essential parts of the car (like in sport cars), it is likely for them to be taken out. Due to the speed, gearbox and engine characteristics can be adapted to reach high efficiency. Furthermore, it is very important to have good acceleration to be able to pull in on highways on the very short Mexican drive-ups. This may be reached by an adaptation of gear ratios. Furthermore, cars in Mexico (and the United States) are normally optimized to a maximum speed of 75 mph (or 120 km/h). The combustion has to be optimal at this speed, and engine characteristics are aligned in accordance (engine application).

The German freeways—so-called Autobahnen—with an overall length of 11,515 km, covering nearly all of the country, are famous for their good quality standards and, even more, for no existing general speed limits. Highways and inner-city streets also are in very good condition. All of the roads are paved.[17] Due to their excellent freeways, Germans really make use of the available maximum speed of their cars. Germans like to go fast on their excellent Autobahn freeway system, up to 250 km/h (most cars are electronically limited to 250 km/h) or even faster (there is no general speed limit on Autobahnen). Of course, German cars need high stability even at this speed. Comfort aligned cars like in America have to be reinforced to be stable. This can be achieved by adjusting springs and dampers (higher stiffness) or reducing ground clearance (lower center of gravity). The gear ratios have to be optimized for a good acceleration from 0 km/h to high speed—different from American cars, which are optimized from 0 to 55 mph (88 km/h). Engines also need to be adapted, and German cars do need high-performance tires, which are able to sustain the high speeds.

The final geographical localization factor is traffic conditions. Almost half of the cars operating in Mexico are found in the valley of Mexico City. Traffic in Mexico City is terrible; there is a lot of stop-and-go traffic nearly all times of the day, and many people have to take 2 h or more to commute to work. Insufficient public transport systems and inner-city expressways add to the very high traffic density. Parking is difficult to find within the city. Traffic conditions in Mexico City have to be considered as well in the car setup. For example, the cooling system and the air conditioning have to be stable with the engine at idle most of the time for several hours, due to the heavy stop-and-go traffic. Crossing the city can easily take more than 3 h. The average working temperature of the clutch will be different in these conditions, thus affecting the wear and tear rate, which is a direct function of this working temperature. Due to the high number of cars in the Mexico City area and the limited space, there is a special need for subcompact cars. Outside of Mexico City, in the more rural areas, sufficient space is available. Mexico is a country with left-hand drive; cars drive on the right side of the street. The traffic density in Germany is very high as well. However, thanks to the many freeways and highways, well-designed traffic influence systems, and good public transport systems, the traffic flows smoothly during most times of the day. Parking space inside cities is rare. During vacation time, traffic jams on the express freeways are frequent, with overall lengths of 1,000 km or more. Germany is a left-hand drive country, too. Table 6.4 summarizes and compares geographical criteria.

Required Factor 2: Legal System

A second set of required factors involves laws governing design, production, and sales. Legal regulations have a long history in the automobile industry, and range from environmental legislation, to security laws, to local content

TABLE 6.4

Summary of Geographical Criteria

Geographical Criteria	Mexico	Germany
Location		
Area	1,972,550 km²	357,021 km²
Population (July 2004 est.)	104,959,594	82,424,609
Climate	*Mexico City*	*Berlin*
Temperature average	16°C	8.9°C
Average high temperature	23°C	13°C
Average low temperature	11°C	5°C
Highest temperature	32°C	35°C
Lowest temperature	–3°C	–23°C
Average morning humidity	79%	86%
Average evening humidity	37%	65%
Altitude above sea level	2,234 m	50 m
Average station pressure	780 hPa	1009 hPa
Rainfalls	634.3 ml/a	580.7 ml/a
Average days with snowfall	0	49
Sunshine hours per annum	2,428 h	1,634 h
Global radiation	6,642 MJ/m²	3,566 MJ/m²
Topography		
Elevation extreme	5,700 m	2,963 m
Average elevation	1,500–2,500 m	0–500 m
Average air density		
Infrastructure		
Highway system total (1999 est.)	329,532 km	230,735 km
Highway system paved (1999 est.)	108,087 km	230,735 km
Expressways (included) (1999 est.)	6,429 km	11,515 km
Traffic		
Driver's position	Left-hand drive	Left-hand drive
Number of cars in use (approx.)	12,300,000	43,000,000
Cars/1,000 inhabitants (2002)	127	541

Source: *The World Factbook 2004* (see note 17).

requirements, to traffic laws. The first legislation included limits on allowable speed and other basic traffic rules, mainly to ensure the safety of all traffic participants and a smooth flow of traffic. In 1965, California enacted the first legislation concerning exhaust emissions. Similar regulations were introduced in the whole United States in 1968. Since the United States was the biggest car market in the world, all car manufacturers had to adapt products to the new regulations. Many countries followed this example and introduced their own environmental and safety legislation. By the late 1970s, engineers all over the world struggled to meet the new requirements, which differed across countries and regions. Today, manufacturers even introduce new safety and environmental equipment on their own initiative, understanding that this generates additional value for their clients.

Environmental protection regulations always had and still have a strong influence on car design. The development of exhaust gas after treatment devices was driven by more and more stringent emission regulations. Mexican emission limits are less strict than European and U.S. limits. Differences of more than 200 or even 300% can be found for emissions such as nitrogen oxides (NOx). Different regulations have to be met for high-altitude markets such as Mexico City. The values are given according to the Federal Test Procedure (FTP) of the United States. The obligation of on-board diagnostics (OBD) was introduced gradually from 2001 until 2005. On-board diagnostic systems for use in both Europe and the United States will eventually be accepted in Mexico.

In addition to meeting legislative mandates, designers also must deal with other complications. For example, Mexico plans to progressively introduce U.S. tier 2 or, as an alternative, the Euro 4 automobile pollution standards between 2006 and 2009. However, these more rigorous standards assume that low-sulfur fuels are available. Mexico's native crude has a high sulfur content, which makes it very expensive to produce U.S.-European-style fuels, and Mexico's government-owned oil company (Pemex) has been slow to provide low-sulfur fuel. However, Pemex also keeps the price of fuel artificially low and stable, making fuel economy less important to Mexican consumers. In addition, trade groups impose requirements on designers. Neither the Mexican nor the German government has imposed legal limits for the emission of CO_2, even though it is an important cause of the ozone depletion and global climate change. However, European manufacturers have voluntarily committed to reducing CO_2 emisions. The German Association of Vehicle Manufacturers led the way when it agreed to reduce average fuel consumption by 25% between 1990 and 2005. Going further, the European Automobile Manufacturers Association (ACEA) has voluntarily committed itself to reduce average CO_2 emissions on all new cars to 140 g/km. As a result, designing vehicles with high fuel efficiency is a more important priority in Germany than it is in Mexico.

The most important adaptations are made to the engine itself. New programming of the engine's control unit is necessary in order to optimize performance in different temperatures, humidity levels, altitudes, and local emission requirements. Exhaust systems and transmissions also have to be modified. Calibration of controllers, exhaust gas after treatment systems, and the transmission has to be done. All variables are interdependent, making this task highly complex and very expensive, requiring extensive engineering knowledge. Similar complications are created by laws regarding noise emissions.

Increasing safety requirements are another very important driver of technological advance and innovation in the automobile industry. Mexico still has not established any consistent legal requirements regarding vehicle safety, but manufacturers have voluntarily added chassis reinforcements, safety belts, and air bags. Their engineers also have developed modern active safety systems such as distance radar and braking assistance. Today engineers are working on better driver assistant systems, such as adaptive headlights or night vision systems, as well as pedestrian security. In contrast, German engineers must fulfill all of the directives of the European Commission as well as more rigorous U.S. crash test requirements if vehicles are to be exported.

Required Factor 3: Technological Influences

Technical conventions and state-of-the-art norms and standards will be presented here as technological influence factors. Regardless of the fact that some of them also are legal obligations, the focus lies more on factors, arising out of technical necessity or standardization needs. As we explained earlier, the crude oil available in Mexico has a much higher sulfur content than the fuel that is available in the United States and Europe. Furthermore, the premium and diesel fuels are not available at every Pemex gas station throughout the country. A special niche product in Mexico is liquefied petroleum gas (LPG). It consists of hydrocarbon gases, which are a mixture of propane and butane, usually with a small propylene and butylene concentration. It is often seen as green fuel, as it decreases exhaust emissions. It is widely available in the Mexico City metropolitan area, as well as in other big cities. Although LPG has a high octane number (RON) of 110, it actually has a lower energy content than octane petrol, resulting in an overall low power output, again creating demands on the design of engines, transmissions, and exhaust systems. In Germany, three different types of gasoline are offered, normal, super, and super plus, as well as diesel (DIN EN 590) and, in many places, alternative fuels like biodiesel or hydrogen. Extremely clean and high-octane gasoline is available, reaching octane numbers of RON 100, and different gasolines are produced during summer and winter seasons.

Even some of the technical requirements for multimedia devices vary across the two countries. Cars with DVD players have to be configured to region code 4. A satellite GPS navigation system is not available. In Germany, like in the rest of Europe and in Africa, a frequency of 868 MHz is prescribed for remote control door lock systems. Frequencies for mobile phones are the Global System for Mobile Communication (GSM) standard frequencies of 900 and 1,800 MHz. DVD players in Germany have to be configured to region code 2. The automobile manufacturer has to guarantee the use of the correct frequencies, according to the country's regulations. Adaptations to the transmitting and receiving device of the remote door lock system have to be ensured, as well as software configuration of multimedia devices.

Engineers also must be aware of differences in the availability and quality of raw materials. Different material norms are imposed by professional associations in the two countries: the Society of American Engineers (SAE) in

Mexico and the German Institute for Standardization (DIN), the International Organization for Standardization (ISO), or the European Standards in Germany. This could make it difficult to find exactly the same raw material or purchased part in Mexico as in Germany. Different laws also exist, such as the European prohibition of heavy metals usage (lead, cadmium, mercury, and chrome VI) (Norm 2000/53/EG). As a result, there are various reasons that make it necessary to change a material in the assembly of a car. Additional testing and often expensive liberation procedures may be required. Cars produced in Mexico, but with export destinations within Europe, are also affected by the European laws and standards.

In sum, every country has a distinctive set of required factors that influence product design, development, and production. In many ways, Germany and Mexico are very different countries, but in other ways they are similar. The challenge of localization is devising strategies for turning the differences into advantages, while maintaining the efficiencies provided by the similarities.

Optional Factor 1: Economic Situation and System

While our list of required factors focused on tangible considerations, localization also must deal with a group of more intangible considerations. The first of these involves the economy of a country, a very complex system covering production, distribution, marketing, and consumption of goods and services. Mexico is an emerging economy, the world's thirteenth largest, but one that is closing the gap with the industrialized countries. Important manufacturing industries are food and beverages, tobacco, chemicals, iron and steel, petroleum, mining (silver, iron), and last but not least, motor vehicles. Mexico is also an important producer of agricultural products, e.g., vegetables like corn and avocados. It is a strong export nation, with the United States as its main trading partner (87.6% of exports), followed by Canada (1.8%) and Germany (1.2%). It is a member of the World Trade Association and the Asia-Pacific Cooperation Forum and has free trade agreements with the United States, Canada, Costa Rica, Bolivia, Venezuela, Colombia, Nicaragua, Guatemala, Honduras, El Salvador, and Chile. It established the first transatlantic free-trade area by signing an agreement with the European Union in 1998. Ninety percent of its trade area is now under free trade agreements. Mexico has the thirteen biggest proven oil reserves in the world and is a major oil exporter, especially to the United States.[18] In contrast, Germany is a highly developed and industrialized country. Important products include automobiles, iron, steel, chemicals, machinery, machine tools, and electronics. In new economies like IT and biotechnology, Germany takes a leading position. It takes second place worldwide in exports, only outperformed by the United States. Its main export destinations are France (10.6%), the United States (9.3%), and the UK (8.4%).[19]

Perhaps the most important economic factor in a society is its level and distribution of income and wealth. Until income exceeds a threshold value, it directly determines the preferences of customers. It is the most important purchase criteria in Mexico. The World Bank classifies Mexico as an upper-middle-income developing country. Its gross national income per capita of US$6,230 is one-fourth that of Germany. The income distribution remains highly unequal. Up to 40% of the Mexican population lives in poverty.[20] Germany has a gross national income per capita of US$25,250.[21] The lower discretionary income in Mexico limits the size of the automobile market and underlies a preference for economy cars. Production of the original VW Beetle for the Mexican national market until 2003 is a good example. The Beetle's successors are the VW Pointer, imported from Brazil, and the Nissan Tsuru, assembled in Mexico. Both are very inexpensive cars with only basic features, outdated technology, high levels of emissions, low levels of safety, but attractive styling, an important factor in the Mexican market.

Mexico's lower level of income also is complicated by the value of its currency. The cost of operating in a global business environment depends on the exchange rate of the country's currency. The national currency is the Mexican peso, and its value "floats" in the international currency market with only limited and indirect control by the Mexican government. The currency of Germany, as of all members of the European Monetary Union, is the euro, which together with the U.S. dollar and the yen, is one of the most important currencies worldwide. Between 2002 and 2005, the value of the peso was continuously depreciating relative to the euro, reaching 15 pesos/euro at the end of 2004. The devalued peso made Mexican products very inexpensive in Europe, thus increasing export sales and profits. Mexican carmakers could earn euros through export to buy more devaluated pesos to pay for local raw material and labor. A weak peso also attracted automobile manufacturers and suppliers to Mexico. However, cars and parts imported from Germany were getting steadily more expensive, making them less competitive. The peso recovered somewhat during 2005, but its value declined again in 2006.

Labor costs and productivity also are important facts for local automobile production. The hourly compensation costs of production workers in Mexican manufacturing were at US$2.48 in 2003.[22] The labor costs in Germany are among the world's highest. Hourly compensation for a production worker in manufacturing is at US$29.91, twelve times Mexico's rate.[23] High wage rates give companies incentives to automate, which increases worker productivity. In contrast, Mexico's low labor costs make manual, labor-intensive work affordable, reducing incentives for companies to invest in automation or take other steps to increase productivity. However, in some cases production volume is so high that companies can justify the investment, amortization, and ongoing maintenance costs involved in the use of robots. Or, where high accuracy or process reliability is needed, or personal safety is at risk, robots have to be used. But, in most cases, labor is preferable to investing in new technology. Productivity, and thus wage rates, stay low, limiting the size of the Mexican market and dictating a particular kind of vehicle.

Optional Factor 2: Sociopolitical Situation and System

Demography is the statistical description of a population, focusing on social and economic dimensions. Key demographic indicators are age and life expectancy, income and wealth, degree of urbanization, and economics. Demographic data are used in the design of products and in marketing a product or service to target customers. Compared to the countries of Western and Northern Europe, Mexico has a young and growing population—its median age is 24.6 years and its annual population growth rate is 1.18%. About one-quarter of Mexicans live in rural areas, where the population density is low: 54 persons/km². Population density in Mexico's cities is very high, especially in the capital, a conglomeration of 22.5 million people with a population density of 5,700 persons/km². Germany is the most populated country in Europe, with a population roughly 80% of Mexico's (84 million vs. 104 million) but a population density five times that of Mexico (231 persons/km²). Almost 90% of Germans live in urban areas, the median age is almost twice Mexico's (42 years), and its population has been stable or declining for the past decade. The most common family living arrangement is a single (or nuclear) family. Fertility rates (the number of children born to each woman) are half Mexico's (1.38 vs. 2.39), and families with more than three children are rare. In contrast, Mexican households often consist of three generations, families with four or more children are common, and children generally live with their parents until they marry.

These differences can have a major effect on product design and development. Young people often have different needs and preferences than older people, and they use their autos in different ways. Germany's smaller families mean that a subcompact car is sufficient for nearly all transportation needs. In Mexico, where families are bigger, there is a demand for cars with more space, although Mexicans' style preferences and the population density of its cities lead to a preference for smaller calls than family size alone would predict.

Germany and Mexico also differ markedly in their education system and the educational attainment of their citizens. An automobile is a technical, very complex product. Its development, construction, and maintenance all require high levels of general education and specialized skills and training. Therefore, workforce education levels and qualifications are an important factor for the automobile industry. Education can be seen "as a determinant of national competitive advantage"[24]; a poor educational system or one that is limited to a small elite handicaps economic development and requires employers to spend a great deal of time and money compensating for it. A 2003 comparative study found that the competence levels of 15-year-old Mexican students in mathematics, reading, and natural science were ranked last of all developed countries.[25] Mexico's government is aware that the poor education of its workforce is a serious barrier to global competitiveness. Therefore, investment of public funds in the country's education system has increased significantly during recent years. Mexico currently invests 5.9% of

TABLE 6.5

Summary of Educational Criteria

Education	Mexico	Germany
Investments in Education		
Percentage of GDP invested in education	5.9%	5.3%
Share of public spending invested in education	24.3%	9.7%
Primary spending	US$1,357	US$4,237
Secondary spending	US$3,144	US$5,366
Tertiary spending	US$4,341	US$10,504
Results		
Literacy (over 15 can read and write)	92.2%	99%
Completion of secondary education	21%	85%
University level attainment	5%	19%
PISA 2003 OECD Results (Avg. Points/Rank)		
Mathematical literacy	385/29 of 29	503/16 of 29
Reading competence	400/29 of 29	491/19 of 29
Natural science competence	405/29 of 29	502/15 of 29

Source: *Learning for Tomorrow's World and Education at a Glance 2003* (see notes 25 and 29).

its gross domestic product (GDP) in education, and spending on education is rising faster than GDP per capita.[26] However, even with these increases, total spending for primary, secondary, and tertiary level education remains far below the average for developed countries. Literacy is at 92.2%, but only 21% of 25- to 34-year-olds have completed an upper secondary level education, and only 5% have college degrees.[27]

Germany spends a slightly lower proportion of its GDP on education (5.3%), but its overall spending is two to three times higher than Mexico's. Literacy in Germany is at 99%. Eighty-five percent of German workers receive post-secondary training, through its unique l-system, which combines upper secondary level education in public schools with on-the-job training in private companies. The university attainment level is 19%, above the average for developed countries, and four times higher than in Mexico.[28] In the 2003 study, Germany's achievement level was slightly above average for developed countries. At this point, German policy makers are more concerned with the potential shortage of trained workers because of its low birthrate than with their education level. A number of potential solutions are being debated, but none has been fully implemented.[29] Table 6.5 summarizes and compares educational criteria.

These differences in education affect users, mechanics, production workers, and engineers. Mexican drivers prefer simple, easy-to-handle technology in their cars. This makes it easier to handle and interact with their cars, and allows them to perform simple maintenance tasks on their own. More complex and demanding automobile technology, especially the use of electric and electronic technology, is a serious challenge even for many of Mexico's trained mechanics. Mexico's independent repair shops are often very small, and they are often family-run businesses. Sons learn the business from their fathers without any professional training. Few own car diagnosis computers. Without them, many of the systems in modern automobiles, like the electrical car network or the electronic control unit of the motor, on-board diagnosis systems, air bag technology, modern braking assistants, and dynamic stability systems, cannot be attended to adequately. This also is true for many comfort features, like electronically controlled air conditioning systems. The repair shops of official dealers often have adequate equipment and training for their technicians, but dealerships often can only be found in big cities, and only a small part of the population is able to pay for this very expensive quality service.

Optional Factor 3: Culture

As we explained in Chapter 2, the culture from which an organization draws its members has a significant effect on its operations—the ways in which people act, make sense out of their actions and the actions of their co-workers, solve problems, and reconcile conflicts and dilemmas.[30] The beliefs, values, and practices that make up a culture often require organizations to adapt their products, production processes, and marketing strategies.[31] Marketing textbooks are filled with examples of errors in cross-cultural operations. A standard example involved the name that General Motors chose for its new Chevrolet Nova when it was exported to Spanish-speaking Puerto Rico. *Nova* literally translated means "star," but when it is spoken it sounds like "no va," which means "It doesn't go." GM had to change the car's name.[32] Ironically, since the English meaning of *nova* is a star that looks bright at first, but quickly fades and dies, it may not have been an especially good name for an automobile in the United States either.

The basis of a society is its language, which both reflects underlying cultural assumptions and influences the ways in which people perceive the world around them and make sense of everyday events. Language influences every part of the operation of a business. Colleagues working together, dealers interacting with clients, marketing with their target groups, and customers learning to operate their vehicles all rely on language and are prone to misunderstandings because of language differences. The most common symbol systems are everyday language and rituals, both secular and religious. Like most former colonies of European countries, Mexico's official language is that of its colonizer—Spain. However, there still exist various Mayan, Nahuatl, and other regional indigenous languages. English is only spoken by a minority of the population. The official language of Germany is its native language, but many immigrant and temporary workers only speak their mother tongue, for example, Turkish. English is taught obligatorily in schools, making English a widely spread second language for Germans.

Religious beliefs and practices are closely linked to cultures and help shape attitudes toward work, risk, power relationships, ethics, and even entrepreneurship.[33] Roman Catholicism was established in Mexico during the Spanish conquest and colonization and currently is the dominant religion (89% of Mexicans are Catholic). Although Mexicans often were coerced into converting, they often retained their previous beliefs and fused them with the new religion. Two famous examples are the common veneration of the *Virgen de Guadalupe* (Our Lady of Guadalupe) and the *Día de los Muertos* (Day of the Dead), in which the European Catholic All Saints Day is combined with indigenous rites of ancestor veneration.

German's religious beliefs are more diverse (about a third are Roman Catholic, a third are Protestant, and a third have no religious affiliation); it is a more secular society than Mexico. In theory, Protestantism, which started in Germany, emphasizes the values of individualism, equality, hard work, self-sacrifice, and planning for the future, values that also have traditionally been associated with capitalist economic systems.[34] This preference for structure and planning seems to explain why people from other cultures often view German product development as mired in overengineering, a term that refers to a tendency to stubbornly follow established designs and rules, sometimes leading to the overfulfillment of requirements. Designs, products, and solutions should work perfectly the first time. It also means that Germans feel most comfortable in structured, no-nonsense meetings and readily accept feedback, negative or positive, as necessary for personal improvement. Punctuality and efficiency are virtues. They always aim to finish projects on time and are very punctual. German companies usually promote employees according to their individual performance. In contrast, Mexican engineering focuses on creative adaptation and "making do" with available resources, and view meetings as interpersonal events that also involve decision making. Catholicism legitimizes top-down control and community connections, especially families, and promotion decisions are based on loyalty, seniority, and personal connections.[35]

Chapter 2 introduced Hofstede's research on the various dimensions of culture. The three factors that are most important to a comparison of German and Mexican culture are power distance, individualism-collectivism, uncertainly avoidance, and masculinity-femininity. Mexico's colonial past left a legacy of high power distance (deference to people of higher social or organizational rank), and today's wide differences in wealth and social class perpetuate it. This is reflected in the hierarchical structure of Mexican organizations, which are very authoritarian. Mexican workers expect and accept clear instructions from their supervisors. In Germany, with a low level of power distance, a more participatory, democratic decision process is expected.

A second difference involves the individualism vs. collectivism dimension. Germany's high individualism value underlies an open and direct communication style as well as a focus on individual achievement and freedom. Mexico is a collectivist society, with the family as the most important organization. Group membership dominates individual opinion. The third dimension, uncertainty avoidance, measures the extent to which members of a society accept ambiguous situations and tolerate uncertainty. Germany and Mexico both had high uncertainty avoidance scores, but the two cultures manage it in different ways. Germans reduce uncertainty through rules and regulations; Mexicans manage it through authority relationships. Both societies also are risk-averse. The two societies also had similar masculinity vs. femininity scores—both are highly masculine, but they have different conceptions of time. For example, Mexican workers find it very difficult to use Kaizen, or any kind of continuous improvement process (CIP). Change creates uncertainty, and processes of continuous change maximize it. CIP requires a long-term view of things, while Mexican culture focuses on living in the moment. CIP also depends on participation, on individuals playing the role of supervisors. While all of these characteristics are consistent with German culture, they violate Mexican cultural expectations. On the other hand, Total Quality Management (TQM), like all quality control tools, does not tolerate any uncertainty, and therefore favors cultures with high uncertainty avoidance, like Mexico and Germany. In production and even more in development, Mexican workers and engineers need extensive training to be able to work in the structures of multinational car manufacturers.

Conclusion

This comparative case study of Mexico and Germany was provided in order to illustrate how key localization factors influence global engineering. Advantages and disadvantages of localization were explained. Impacts of localization on product development and production processes have been analyzed. The interdependency with mass customization and build-to-order models has been assessed. The main strategic options for car manufacturers concerning their business, design, and production strategy have been analyzed. Localization will remain an important issue for the automobile industry throughout the foreseeable future. Common test procedures and certification requirements of evaporative and exhaust emissions, as well as safety standards, could greatly reduce the development and certification test costs, and therefore help to avoid non-value-adding localization to different legal environments. Voluntary localization could greatly improve competitiveness of a car in foreign markets, without offering high-level customization possibilities. This would help to stay in control of the rising number of variants and the resulting high complexity. As the globalization process continues, the importance of localization will rise accordingly. Further investigation efforts should include a more detailed study on the impacts of localization on complexity in production processes.

Review and Study Questions

1. Define *internationalization, globalization, localization, glocalization,* and *modularization.*

2. Discuss the differences among:
   a. Cultural neutral products
   b. Homologation
   c. Customization

3. Why is automotive and parts production concentrating in a few manufacturers?

4. Explain the main differences between German and Mexican automotive production.

5. How will the future challenges in automotive production affect the German, Mexican, American, etc., automotive production.

6. Explain how the safety regulations in your country affect automotive localization.

7. Find materials availability in your country different from those in other big automotive producer countries.

8. Analyze the culture model of your country and contrast it with Germany's and Mexico's.

9. Select one premium and one economic car in your country and investigate the differences in localization.

10. Investigate the car development process in your country and point out the localization activities done during that process.

11. Propose a sound product, manufacturing, and business strategy for car production.

12. Similar to the automotive product localization, give the factors, subfactors, and impacts on product design (different from an automobile) for the Mexican, American, European, Asian, etc., market. An example list of products to choose from is television sets, cell phones, bicycles, motorcycles, software, refrigerators, computers, and so on.

Name of product to analyze: _____

Factor	Subfactors	Explanation of the Impact on Product Design
1. Geography	1a. Topography ____	
	1b. Weather ____	
	1c. ____	
2.	2a. ____	
	2b. ____	
	2c. ____	

Name of product to analyze: _____

Factor	Subfactors	Explanation of the Impact on Product Design
3.	3a. _____	
	3b. _____	
	3c. _____	
4. Cultural	4a. _____	
	4b. _____	
	4c. _____	

13. Active exercise: *Individually,* discover car design features that are different, depending on the automaker origin and region application. *As a team,* compare your lists and compile a list that represents the consensus for the team.

Design Feature	American Car	European Car	Mexican Car

14. Active exercise: *Individually,* discover car design features affected by optional localization factors for the Mexican market. *As a team,* compare your lists and compile a list that represents the consensus for the team.

Optional Localization Factor	Explain Design Feature
Economical	
Fuel price	
Average income	
Sociopolitical	
Personal safety	
Governmental policies	
Cultural	
Traditions	
Values	

15. For a product (different from a car) give two localization examples and two customization examples. Give reasons.

Name of product: _____

	Adaptation Reason	Adaptation Type
1. Localization	1.1. _____	
	1.2. _____	
2. Customization	2.1. _____	
	2.2. _____	

Notes

1. Timothy Sturgeon and Richard Florida, *The Globalization of Automobile Production*, research note for the International Motor Vehicle Program Policy Forum (South Korea, 1997).
2. Jan Dannenberg and Christian Kleinhaus, "The Coming Age of Collaboration in the Automotive Industry," *Mercer Management Journal* 17 (2004): 88–94.
3. Kelsey Mays, "A Closer Look at Domestic-Parts Content," Cars.com, http://www.cars.com/go/advice/Story.jsp?section=top&story=amMadeParts&subject=ami (accessed December 17, 2008).
4. George Will, "Despite Cuts, Ford's Fate Tied to Other U.S. Automakers," *Houston Chronicle*, December 18, 2008, p. B9.
5. Mohan Subramaniam and Keoy Hewett, "Balancing Standardization and Adaptation for Product Performance in International Markets," *Management International Review* 44 (2004): 171–194.
6. R. J. Calantone, S. T. Cavusgil, J. B. Schmidt, and G.-C. Shin, "Internationalization and the Dynamics of Product Adaptation: An Empirical Investigation," *Journal of Product Innovation Management* 21 (2004): 185–198.
7. Felix Rauner, "Localization: The Transfer of Cultural Neutral Technologies into Cultural Applied Products," in *Proceedings of International Institute of Manufacturing and Culture* (Nagoya, Japan: 2002).
8. Takao Enkawa and Shane Schvaneveldt, "Just-in-Time, Lean Production, and Complementary Paradigms," in *Handbook of Industrial Engineering: Technology and Operation Management*, 3rd ed., ed. Gavriel Salvendy (Cambridge, UK: Cambridge University Press, 2001), pp. 544–561.
9. Mitchell M. Tseng and Jianxin Jiao, "Mass Customization," in Salvendy, ed., *Handbook of Industrial Engineering*, pp. 684–709.
10. Oliver Wyman, *2015 Car Innovation Report* (2007), http://www.oliverwyman.com/de/pdf-files/Car_Innovation_2015_deutsch.pdf.
11. Ibid.
12. Alexandra Olson, "Troubles in U.S. Felt across the Border," *Houston Chronicle*, December 26, 2008, p. D3.

13. Juan Carlos Moreno, *Mexico's Auto Industry after NAFTA: A Successful Experience in Restructuring?* Kellogg Institute Working Paper 232 (Cambridge, MA: Rockefeller Center for Latin American Studies, Harvard University, 1996).

14. *Automobile Manufacturers in Germany: Industry Profile* (New York: Datamonitor, 2004).

15. Ibid.

16. Nikolai V. Lyachenkov and Michail P. Ramasanov, "Adaptation of the Passenger Car to the Road Conditions of Third World Countries," paper presented at the Environmental Sustainability Conference and Exhibition (Graz, Austria: Austria Society of Automotive Engineers, 2001).

17. Central Intelligence Agency, *The World Factbook 2004* (Washington, DC: U.S. Government Printing House, 2004).

18. Ibid.

19. Ibid.

20. Ibid.

21. The World Bank, *World Development Indicators 2004* (New York: The World Bank, September 2004).

22. U.S. Department of Labor, Bureau of Labor Statistics, *International Comparison of Hourly Compensation Costs for Production Workers in Manufacturing*, November 2004.

23. Ibid.

24. Michael E. Porter, *The Competitive Advantage of Nations* (New York: Free Press, 1990).

25. Organisation for Economic Co-operation and Development (OECD), *Education at a Glance 2003* (Paris: OECD, 2004).

26. Ibid.

27. CIA, *World Fact Book*.

28. OECD, *Education at a Glance 2003*.

29. Organisation for Economic Co-operation and Development (OECD), *Learning for Tomorrow's World: First Results from PISA 2003* (Paris: OECD, 2004).

30. Fons Trompenaars and Charles Hampden-Turner, *Riding the Waves of Culture: Understanding Cultural Diversity in Global Business*, 2nd ed. (New York: McGraw-Hill, 1998).

31. Charles W. Hill, *International Business: Competing in the Global Marketplace*, 3rd int. ed. (New York: McGraw-Hill, 2001).

32. D. A. Ricks, *Big Business Blunders: Mistakes in International Marketing* (Homewood IL: Dow Jones-Irwin, 1983).

33. Hill, *International Business*.

34. Hill, *International Business*; Richard H. Tawney, *Religion and the Rise of Capitalism* (Gloucester, MA: P. Smith, 1962); Max Weber, *The Protestant Ethic and the Spirit of Capitalism* (New York: Scribners and Sons, 1958).

35. María Eugenia de la O et al., ed., *Los estudios sobre la cultura obrera en México* (Mexico City: Conaculta, 1997); Rocío Guadarrama Olvera, *Cultura y trabajo en México* (Mexico City: UAM, 1988); Carlos Miller, "Indigenous People Wouldn't Let 'Day of the Dead' Die," *The Arizona Republic*, 2004, http://www.azcentral.com/ent/dead/history/.

Section III

Case Studies: Engineering Predominance

7

SmartDrill Stays Home

SmartDrill is a U.S. manufacturing company that was founded in the late 1970s when the energy crisis created an increased demand for oil exploration. Its main products are bits designed for oil and natural gas drilling. Operating for decades, SmartDrill has been regarded as a world leader in drill bit manufacturing. Although it operates in several countries, there are three main plants producing cutting bits, one in Houston, Texas, one in England, and another in Singapore. The production of drill bits in these three plants accounts for more than half of the drill bits consumed worldwide. The company's home office is in Houston, and its plant in Houston is a center for producing bits to be used in the Americas. The plant in England serves European customers, and the Singapore plant is a hub for the Asia/Pacific region. Figure 7.1 illustrates a finished product.

SmartDrill's core facility is the Houston plant, which is divided into three primary areas: machine shop, heat treatment, and assembly areas. Other parts of the plant produce subcomponents used in the core functions. These departments include hard metals, a forge shop, and power metal cutter (PMC) operations. PMC involves a patented process for making rock bit cutters from powered steel using a combination of powder metal technology and forging technology. To produce the highly complex drill bits made in the Houston plant, several activities must be completed and coordinated, including a very challenging and labor-intensive welding task. During the past 4 years, SmartDrill has improved the product quality and plant productivity by investing more than $6 million to upgrade and automate its core facility. The Houston plant employs slightly more than four hundred employees, operating in approximately 460,000 ft^2 of covered space in eight major buildings.

With the high-technology equipment, work performed in the Houston plant usually requires the use of highly skilled workers. Some of the skills required of the workers are to operate Computer Numerical Control (CNC) programming and a three-dimensional precision measurement. The drill bit manufacturing diagram is demonstrated in Figure 7.2.

SmartDrill obtains components and materials for the Houston plant from the company's operations in a number of different countries. Houston's supply chain scheme is shown in Figure 7.3. A Canadian supplier provides steel powder for PMC operation. Tungsten carbide is produced in China and imported to Houston by a U.S. trader. Production of an insert for the bits usually is outsourced to other companies in the United States, but a highly specialized diamond-coated insert one is produced at SmartDrill's England plant.

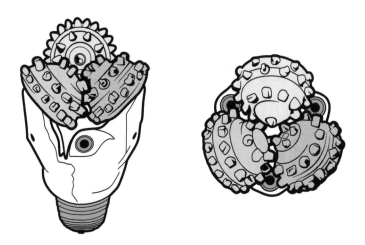

FIGURE 7.1
Completed drill bit.

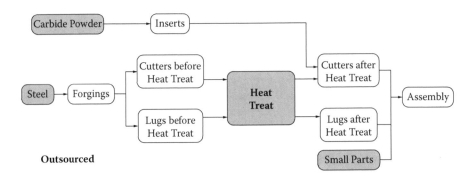

FIGURE 7.2
Drilling bit manufacturing diagram.

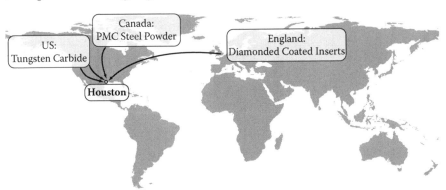

FIGURE 7.3
Houston plant's supply chain.

In addition to the Houston facility, SmartDrill established a similar operation plant in Singapore in 1980. The Singapore plant operates in a 60,000 ft^2 facility and employs nearly two hundred employees. In 2002, Singapore outperformed Houston with bits produced per day (sixty vs. thirty-five). However, most of the machines in the Singapore facility are outdated and unable to respond to customers' demands for more sophisticated bit patterns that require higher levels of precision than the equipment in Singapore can provide. As a result, SmartDrill has currently allocated 60% of its total drill bit production to Houston and 40% to Singapore. The supply chain for the Singapore plant also involves operations in many countries. The Houston plant supplies Singapore with carbide inserts and raw forging material. As illustrated in Figure 7.4, bushings, seals, ball/roller bearings, thrust washers, and grease and larger forgings are outsourced within the United States and shipped to Singapore. Castings are supplied from SmartDrill's England plant, and hard-facing parts of the bits come from a vendor in Mexico.

In sum, SmartDrill seems to have adopted a production strategy that optimizes its international operations. Products that require specialized, high-level engineering are designed in the Houston facility, whose employees possess the requisite expertise and whose technology is sufficiently sophisticated to produce complex designs. The components and materials needed for this production come from low-cost, high-quality sources. More routine products are made in Singapore, where costs are lower and the level of technical expertise and equipment is adequate. The Singapore plant does employ engineers, but their primary job is to adapt existing products to the needs of the Asian market. They are not responsible for the high-technology engineering tasks performed in Houston.

However, like every other multinational firm operating in today's global environment, SmartDrill faces constant competitive pressures. As a result, it regularly explores alternative production scenarios that would reduce its

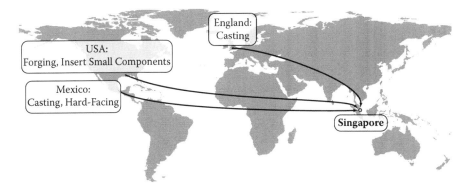

FIGURE 7.4
Singapore plant's supply chain.

overall production costs, allow it to become even more competitive, and gain more market share. In 2002 an outside consulting firm hired by SmartDrill recommended that the organization fundamentally change its way of operating. The process through which the company made decisions about that recommendation is the focus of this case study.

Culture, Communication, and Decision Making

One of the most controversial parts of Hofstede's research on culture was his 1980 argument that the organizational theories developed in the United States may not apply at all to organizations in other cultures, or may apply only if major changes are made that better adapt them to the different values, institutions, and traditions of those cultures. One of the most important of these distinctively U.S. perspectives is the view that organizational decision making can, and should, be strictly rational.[1] The underlying assumptions of this rational actor model are quite simple:

1. An employee recognizes that a problem exists and that it is caused by some unexpected or as yet untreated change in the organization's environment or by the actions of some of its members.
2. Each member of the organization who, because of his or her formal position, expertise, or available information, has an interest in the problem is told about it and invited to help solve it.
3. Alternative courses of action are compared through open, problem-solving communication. All relevant and necessary information is gathered, made available to all decision makers, and objectively assessed.
4. The optimal solution is chosen and implemented.
5. The impact of the decision is monitored and information about its effects is gathered and stored for use in similar situations in the future. Through this feedback process the rational decision process is able to correct itself.[2]

In some cases employees in U.S. firms can and do make decisions in this way. But, there is a substantial amount of evidence that purely rational decisions really are quite rare. A number of factors and pressures lead decision makers to make nonrational decisions, that is, decisions that do not follow the dictates of the rational actor model. This does not mean that their decisions are irrational, that they are blindly emotional or in some way ignore the reality of the situation and problem the decision makers confront. It merely suggests that the assumptions of the rational actor model are unrealistic when applied to many real-world decisions.

The simplest and most important limit to rational decision making is time. Organizational decision makers are often required to act quickly, especially in today's highly dynamic and competitive business environments. Seeking out adequate and accurate information and devising and considering multiple alternatives simply cannot be done. Markus Vodosek and Kathleen Sutcliffe succinctly summarize the dilemma that managers face: "Although in some cases extensive [rational] analyses may lead to better decision making ... it consumes valuable time and resources, decreases the speed with which decisions can be made, and creates a false sense of security."[3] Instead of trying to make rational choices when it is impossible to do so, decision makers play hunches that are guided by their past experiences. Without experience, decision situations are just random bits of information, but with experience, decision makers often can quickly intuit a productive course of action. Herbert Simon, whose research on nonrational aspects of decision making won him the Nobel Prize, used the following example. Present a chess champion and a chess novice a board with twenty-five pieces arranged at random. Let them study the board. Then, remove the pieces and ask the players to replace them in their correct positions. Both players will be able to replace about six pieces accurately. Later on, play a game of chess until there are about twenty-five pieces left on the board. Repeat the experiment. The novice will still be able to replace about six pieces, but the champion will correctly reposition twenty-three or twenty-four pieces. The difference lies in the champion's experience and the way that experience enables the recognition of meaningful patterns. When the pieces are randomly arranged, there are no patterns and the champion's experience does not help. Playing a game creates familiar patterns, which the champion can recognize instantly and intuitively. Organizational decision makers' past experience has the same effect. When confronted with a problem, they draw on that experience, recognize relevant patterns, and recall solutions that worked in the past.[4]

Of course, intuition is not foolproof: many situations only appear to be like past situations, and rapid recognition can be wrong recognition. But, in many situations that confront organizational decision makers, it is more important to *act* than to take the *best* action; accurate perceptions and decisions are nice, but often they are simply unnecessary. Acting also generates new information and encourages employees to communicate with one another in ways that correct misconceptions. As a result, organizations that prefer taking action over strictly rational decision-making processes tend to understand the environmental pressures they face better, are able to update their information more rapidly, and do a better job of adapting quickly to future changes.[5]

However, time limits rational decision making in another way. Some decision situations involve choices that have immediate, easily determined effects. In other cases, a decision maker may not know the impact of his or her decision for a very long period of time. Because situations change rapidly in modern organizations, decisions that have long-term effects are much more risky than ones that have immediate effects. The information needed to make long-term projections is less likely to be readily available and is more likely to be ambiguous or incomplete than the information needed for short-term decisions. As a result, decision makers tend to artificially simplify the situations they face by managing them incrementally. They muddle through complex problems by making a series of small decisions. Eventually, each of these minor decisions provides them with new information and helps them make sense out of the complex situations they face.[6]

Another factor that limits decision makers' ability to be rational actors is organizational power and politics. One of the most important sources of organizational power is information. But, information can be transformed into power only if it is scarce (that is, not available to just anyone) and only if an employee knows how to use it strategically. As a result, the search for relevant and accurate information that is the core of the rational actor model often is much more difficult than it first appears. In theory, all employees have the same incentives to help their organizations make good decisions. But, in practice, revealing information may reduce an employee's power and increase his or her vulnerability; withholding that information can increase power and make an employee more secure. In addition, powerful employees can push an issue through the decision process rapidly or can interrupt the process by pressuring for a longer information search, demanding that other interested parties be involved in the process, tabling the issue, or referring it to a subcommittee. For example, the president of a subsidiary of a large multinational corporation chairs an eleven-person committee that includes the vice presidents and department heads. The group must decide between the terms of an existing contract and a new pricing system. After a half an hour it becomes clear that the president and executive vice president disagree on the proposal. One senior vice president adds information about international market conditions, but no other members speak up, because they realize that doing so may alienate one of the two top-ranking people in the organization. No action is taken, but another meeting is scheduled to discuss the issue further (and then another, and another, etc.).[7]

Finally, decision situations may simply be too complex for decision makers to use rational processes. A group of cognitive biases tend to hamper human information processing and undermine rational decision making. For example, human beings seem to overestimate the likelihood of good outcomes and underestimate the probability of bad ones. For example, no matter how much information students are given about past patterns of grading in a course and about their own academic records, they invariably seem to overestimate their chances of receiving A's and B's and underestimate their chances of getting C's, D's, or F's. There is also a tendency to use information based on its ready availability more than on its quality, or to interpret a decision situation in terms of the past experiences we remember most easily rather than the ones that are more relevant (this is called the availability bias). For instance, a manager trying to decide whether employees will adopt an innovation may call to mind a recent event in which several administrative assistants sabotaged a new word processing program because they wanted to stick with their old word processing application. With this example in mind, the manager is likely to conclude that resistance is very likely. However, the manager's judgment is based on one example; it may well be the case that in making the judgment based on the one example, the manager has overlooked many other cases in which employees embraced innovations.

As a result of these limitations, it is best to think of organizational decision situations as points along a continuum. At one extreme are very simple, recurring problems that are not politically charged. The information that is needed to make the decision is readily available and easy to interpret, the decision is not politically charged, and the impact of the decision is immediate and unambiguous. The organization can establish policies and procedures that cover these situations, and teach its employees how to apply those established policies and procedures to concrete cases. Fortunately, the vast majority of organizational decisions are of this type. At the other extreme are decisions that are so complex that they overwhelm rational processes. The relevant information is unknown, unavailable, or inherently ambiguous. There are an infinite number of possible courses of action, or none. Organizational politics are dominant, and there seem to be no relevant precedents in the organization's or decision makers' past experiences. Neither rational nor intuitive decision making is possible. The demand is to do *something*, not to do the best thing. Between these poles are decisions that are difficult, but manageable.

The Facts of the Case

The consulting team identified two areas of potential improvement. One involved the kind of communication problems that are inevitable between operations that are geographically dispersed. For example, John is a design manager at the Houston plant. He has been working with SmartDrill for more than 20 years. His job is to design and estimate the cost of the drill bits after receiving the market requirements from the marketing department. With the nature of his job, John acts as the key player in the production department (because his blueprints determine how complex the manufacturing will be) and the center of new product information, which has to handle the issues from marketing and customer service departments. Those issues often go far beyond the borders of the United States because both marketing requests and customer service problems can come from anywhere in the world. John often felt frustrated with the operation by people from Singapore because it often took days or weeks to receive a response from them, even to simple questions. Without data from Singapore

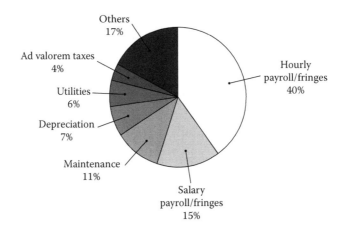

FIGURE 7.5
Distribution of annual expenses for Houston plant in 2001.

he could not finalize product cost estimates. The marketing department was always on his case because they were expecting information from him, but John couldn't get the Singapore people to do anything. The marketing people did not bear in mind that at Singapore it was dinnertime when the Houston plant was barely starting work in the morning. On the other hand, the people from Singapore also had their problems with the Houston plant. John's contacts in Singapore were always busy writing reports, and they did not have as many resources available to do all the requests. Yet when the Singapore plant needed advice to resolve unexpected quality problems or machine setup problems, John was unavailable to help because of the time difference.

The second problem identified by the consultants involved lower productivity at the Houston plant. Although there could be many explanations for the difference, including the increased complexity of the products being designed and produced in the United States, the consultants attributed the difference to the Houston plant's older, highly unionized, and more expensive workforce. The Houston plant employs 260 hourly floor workers, 55 shop supervisors and manufacturing engineers, and 45 additional engineers who work in the material laboratories and R&D department. The total expenses of the Houston plant in 2001 can be broken down as shown in Figure 7.5.

Moving South (and East)

Based on its analysis of the problems that SmartDrill faced, the consulting team recommended that it shut down its Singapore and Houston plants and move its operations to Mexico. Bob, a vice president of the facility in Houston, became the key person leading the feasibility study of the plan. According to his initial research, SmartDrill-Houston spends $26.5 million (M) every year:

$14.7 M = compensation, salary
$4.8 M = salaried employees ($40/h, including engineers)
$6.8 M = direct worker ($16/h)
$3.1 M = indirect worker ($18/h)
$11.8 M = electricity, utilities, tools, purchasing

In Mexico the company could expect to spend $2 million annually:

$6/h = salaried employees (including top-notch engineers)
$2/h = direct worker
$3/h = indirect worker

At first glance, it is obvious that labor costs could be reduced substantially (approximately US$12 million per year) if the plants were moved to Mexico. In addition, the problems related to working in two widely separated time zones would be eliminated. The consulting firm recommended that the shift take place in three stages:

Stage 1: In the first year they would close certain departments to begin out-sourcing. Bob would review which components were purchased versus built in-house. Noncore competency operations would be reviewed as candidates for outsourcing. As an example, machine maintenance or "cutters before heat treat" might be more economical to outsource. Mexican sources for the identified items would be sought.

Stage 2: After the second year SmartDrill would close the Singapore plant and open the one in Mexico. They would have to produce cutters in Mexico and ship the parts to navigation (one of their plants in Houston). Navigation would continue to produce lugs and assemble bits for shipment. This is a relatively low-risk strategy to begin the process of reducing product cost. Each input in the supply chain for the Singapore plant would be shortened and simplified by moving to Mexico. It would take longer to ship completed bits to the Asian market, but those delays would be offset somewhat by increased speed in responding to marketing requests. The complex tasks performed in Houston would still be performed there. Once Mexico was producing cutters, Singapore would be closed.

Stage 3: In the third year SmartDrill would close the remaining operations in the Houston plants and allow the new Mexican plant to set up a focused factory with cellular manufacturing. Additional production would be phased in over time.

Of course, making a change of this magnitude would not be easy. Although the labor cost comparison was easy to calculate, estimating the total costs for the proposed plan would be much more difficult. A team would have to be formed to develop the estimates and construct a timeline for the project. It would have to clearly define its task—to make an economic evaluation of the proposed special facility and plant consolidation activities—and gather information on suppliers, closing down old facilities, and building new ones, determining which activities should be completed in the United States and which should be shifted to Mexico, establishing storage and transportation facilities and networks, and so on. Logistical issues would need to be resolved. Methods of transferring information between plants would need to be considered. Political issues such as determining the appropriate government agencies to contact and proper procedures to follow regarding plant closings and openings in each country would need to be dealt with. Legal issues would also exist regarding ending some contracts and developing new ones, dealing with the union in Houston, and handling employee dismissal, retention, and compensation during the transition period.

Eventually, and in spite of the marked differences in labor costs, SmartDrill decided not to follow the consultants' recommendations. The process they used to make that decision is the most important part of our story.

Why Wouldn't You Move to Mexico?

Understanding SmartDrill's decision making starts with a primer on the Mexican *maquiladora* industry. The term comes from an old Spanish word for the fee that farmers pay a miller to grind grain into flour. Today it refers to a Mexican assembly or manufacturing operation that can be wholly or partially owned and managed by a non-Mexican company. Founded in 1965, the *maquiladora* industry was designed to encourage foreign investors to establish manufacturing plants in northern states of Mexico by offering several cost advantages, such as zero tariffs on imported raw material and machines used to process them.[8] A *maquiladora* uses less expensive Mexican labor to assemble, process, or perform manufacturing operations. Finished or semi-finished products are exported back to the country of origin (or another country) with duties being paid on value added to the original components, or duty-free for North American Free Trade Agreement (NAFTA) countries (Canada and the United States). It is one of Mexico's primary sources of production for exports and domestic employment. As a result, a large consulting industry, supported directly or indirectly by the local, state, and federal government of Mexico, quickly arose that had the task of persuading the United States (and other multinationals) to transfer some of their operations to *maquiladoras.*[9]

Companies choose to operate in Mexico for two primary reasons. First, Mexico offers a cheap labor force, in terms of both wage rates and benefits. Second, NAFTA became effective on January 1, 1994. Not only did NAFTA increase the export of Mexican products to the U.S. and Canadian markets, but it also helped provide safety and certainty to U.S. investors in Mexico. As a result, foreign direct investment in Mexico increased dramatically, reaching more than $10 billion per annum in 1994 and 1997, and more than $7 billion in the recessionary years of 1995 and 1996. Although three-quarters of *maquiladoras* are located in Mexico's six states that share a border with the United States, the industry also has moved south. For instance, 146 *maquiladoras* started between 1994 and 1998 in the state of Puebla, southeast of Mexico City. The *maquiladora* industry accounts for nearly half of Mexico's exports and is primarily responsible for Mexico's economic growth. Nevertheless, according to Martinez-Vazquez and Chen, the net effect of NAFTA on the fast growth of Mexican economy, particularly in the *maquiladora* sector, is unclear. Several other factors, such as the liberalization of foreign trade and investment in Mexico, joining the General Agreement on Tariffs and Trade (GATT), and the collapse of the peso, also contributed to that growth.[10]

Bob's initial investigations were consistent with the consultants' report—labor costs were much lower, especially for engineers, there would be no duties on exports to the United States and Canada, and the production would be much closer to the engineering headquarters in Houston. They would have to outsource their noncore competencies, such as forgings, hard metals, and small parts. But, the initial findings warranted further investigation and a trip to the border.

Table 7.1 provides typical wages per day for varied skill operators in three different cities in Mexico.

Once he started investigating wage rates in more detail, Bob found that the $2 per hour operators promised by the consultants were capable of only low-skill jobs, such as assembling windshield wipers. Workers performing these tasks could be trained within an hour. But, these workers tended to be young, low skill, and had no loyalty to a particular company. They were ready to leave at any time and move to a place that would pay them a few cents more per hour, or once they had earned enough money to meet a specific goal (e.g., to purchase a stereo). Sometimes they would quit if the company across the street had a better lunch menu. Although 80% of the workers that SmartDrill would employ in Mexico would be performing routine tasks, those tasks were much more complex than those performed by these workers. Bob found that to perform the tasks currently being performed in Singapore, SmartDrill would have to recruit skilled workers and provide a moderate level of training. In order to get that level, the company would have to invest as much in training and compensation as it was paying in Singapore; in some cases, the training and compensation costs would equal those of the Houston plant. Furthermore, once workers were trained to that level, they would find it easy to immigrate to the United States and obtain higher-paying jobs. The remaining 20% of the workforce would have to be highly skilled. In Mexico, there is a very limited set of high-skill workers, and they are able to demand wages at least three times those of production workers in order to stay with a particular company. And, like the production workers, once they are trained, they are very likely to try to immigrate to the United States.

TABLE 7.1

Wages per Day in Three Different Cities in Mexico (in USD)

City	Operator				Plant Manager		Production Manager		Production Engineer		Technical Production		Account	
	Without Experience		With Experience											
	Min	Max	Min	Max	Min	Max	Min	Max	Min	Max	Min	Max	Min	Max
Monclova	5.6	6.47	6.42	8.2	104.2	140	63.33	78.3	35.83	41.7	20.4	24.6	11.96	15
Nuevo Laredo	11.44	17	18.64	26	104.2	160	50	66.7	14.6	38.3	10	16.7	18.33	37.5
Reynosa	10.4	15.2	14.4	18	158.3	192	126.7	130	86	90	37	43.2	83.33	90

Source: Adapted from *Costes Industriales en Mexico*, 2008 (www.promexico.gob.mx).

Note: Average wages per day plus benefits and taxes. Journey per 8 hours. Minimum wage per day: US$4.07. Change kind: US$9.40.

In addition, there always is a learning curve involved in starting operations in another country. As earlier chapters pointed out, this is why greenfield starts tend to be more successful than acquisitions. Bob found that successful *maquiladoras* tended to make low-end products requiring very low levels of worker skills, and have had years of experience in Mexico. "They long passed over the learning curve," Bob said.

SmartDrill's situation was different in two aspects. First, in comparison to the other plants, Bob needed more skilled workers, which are, most likely, not available in the *maquila* industry. Secondly, Bob understood that learning to operate in a place with such great differences in social, cultural, legal, and political environment could be costly and risky. He admitted that some companies had successfully passed through the intercultural learning curve and managed to survive. Since only about 10% of *maquiladoras* are efficient and profitable by U.S. standards, Bob concluded that the move was much riskier than the consultants' report had suggested.[11]

A Meeting with Jorge

In order to get more information, Bob met with the CEO of the Mexico firm to which SmartDrill had subcontracted some of its operations. A third-generation Mexican entrepreneur, Jorge openly discussed all the complexities of operating in the Mexican culture as well as the social and legal environment in Mexico. At the outset, Jorge described the *maquiladora* labor force. According to him, approximately 40% of Mexicans are under 20 years of age. These are motivated people, very eager to learn and work. At the same time, Mexicans prefer a slow pace of work. Mexican workers lack the "Protestant work ethic" characteristic of U.S. or Japanese workers. Their loyalties are to people rather than to organizations.

According to Jorge, in order to open a plant in Mexico, a company needs to have enough size and capital as well as a good business plan. To be successful in *maquiladoras*, a company's operation must be low labor-cost-intensive or mass labor driven, such as textiles, agriculture, or mining. More importantly, if a U.S. company wants to go to Mexico, management must first address the human factors in terms of the difference of cultures. Culture shock needs to be dealt with and the effort should be mutual, meaning both American executives and Mexican workers are expected to adjust interculturally. For example, Bob was accustomed to dealing with unions in the United States. But, Jorge explained that recruiting *maquiladora* workers was controlled by a completely unregulated labor mafia that extracted additional costs from potential employers.

Another important point that Jorge stressed was that before training Mexican workers into a particular job, management needs to instill the philosophy of what it is to hold a job. Companies need to have a good selection plan and a good education plan as well. His company is picky in selection. However, once employees are assimilated into the company culture, they tend to stay for a long time. Jorge also mentioned that having senior workers in the company was invaluable because they served as role models for other younger or inexperienced workers. Jorge also noted that his company is not operated like most *maquiladoras*. Workers learn to be fair and respectful to everyone, keep developing new skills, and learn to properly document things. Jorge gives his employees incentive pay for developing new skills. However, Bob found that the concept of incentive pay is a taboo topic among *maquila* owners. Jorge also briefly touched upon other complexities about operating in Mexico. For instance, he warned that firms must stay away from legal troubles. According to him, the nature of Mexican labor law determines, "If you start in trouble, you will have a lot of it." In addition, consultation services (legal and managerial) are available, but "Mexican consultants, especially those representing big consulting companies, are not trustworthy." In spite of all these complexities, Jorge believed that Bob should have gone to Mexico. In fact, after Bob made the decision not to move, Jorge said, "I don't understand why you didn't."

However, there were clues from Bob's meeting with Jorge. Although Bob never explicitly stressed the social/cultural risks involved in operating in a different country, there were several times when Bob showed his hesitation about opening a plant in an unfamiliar environment. For example, when Jorge mentioned that it is difficult to teach Mexicans the American way of doing things, Bob responded, "It is even more difficult to teach the Americans the Mexican way of doing things." When Jorge tried to tell him how to manage the Mexican workers, Bob kept saying, "You're the third generation," "You know their things," etc. All these comments indicated that Bob's reluctance stemmed from a general feeling of insecurity from the risks involved in learning to manage foreign workers and complex operations in an alien culture.

There may have been another emotion-related reason for rejecting the move. While Bob visited Mexico, he observed the squalid living conditions of the *maquiladora* workers. When he met with our research team, he vividly described seeing fifty workers being packed into a fifteen-passenger van—an unbelievably inhuman and unsafe scene to the eyes of an upper-middle-class American. Local leaders explained to him that those living conditions were an improvement for the workers, and that having transportation to and from work was an advantage for them. Bob admitted that he could imagine the differences that the van had brought to the workers' lives; however, he *felt* that he would rather stay away from the scene. When cultural shocks like these are not transcended, it would be very difficult to talk about mutual adaptation.

After a series of interviews, site visits, and detailed investigation, however, the original plan was cancelled, and according to Bob, the company has no plan to reconsider moving to Mexico in the short or even long term (3 to 5 years). Instead, they have decided to outsource some of the simplest production functions to a Mexican company. Bob said the decision was mainly due to economic considerations—the low ratio of skilled to unskilled workers in

Mexico, the high turnover rate, the unexpected level of wages and benefits necessary to retain skilled workers (roughly three times the typical *maquiladora* compensation package), the fact that the *maquiladoras* work for repetitive assembly instead of high skill manufacturing, and the lack of infrastructure to support heavy manufacturing.

Interpretation

The SmartDrill decision case provides an excellent example of how the three levels of our model influence one another (Figure 7.6). SmartDrill's global vision (part of the center circle) has been to develop long-term relationships with suppliers and customers in other cultures so that it can strategically adapt its products and production processes to regional markets. In implementing that vision it has established an implementation system (the middle ring of the model) that uses suppliers throughout the world to provide raw materials and components of the highest quality and lowest costs. Its operations draw on a truly global set of resources to meet the needs of a global market, through a supply chain that efficiently connects raw materials, engineering expertise, and customer demands. The decision about opening a *maquiladora* was completely consistent with the company's global vision and operational system. It chose not to move

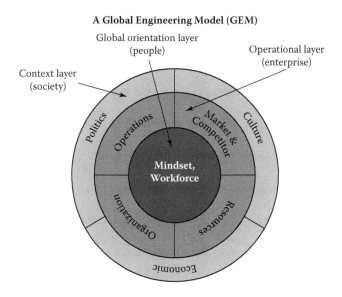

FIGURE 7.6
The global engineering model.

to Mexico because doing so simply did not make sense given its vision and operations.

At the same time, the case study also illustrates how broader considerations (the outer ring of the model) guide and constrain an organization's operations and vision. Detailed economic analyses and careful considerations of the political implications of making the move both explicitly entered into the decision-making process. As important, the company was able to grapple successfully with a core aspect of Western, and especially U.S., cultures—the myth that individuals and organizations can, always should, and usually do make decisions through a strictly rational process, unimpeded by emotional considerations.

Organizational research during the last half of the twentieth century consistently found that this myth was both unrealistic and often counterproductive. Limits to cognitive ability and timely access to relevant and accurate information combine with decision complexity to severely restrict the range of situations in which rational processes are possible or preferable. Conversely, there is increasing evidence that emotions actually can increase decision rationality by helping decision makers make judgments about what is worth thinking about and what information is relevant or irrelevant for problem solving. Emotions also focus decision makers' attention, helping them sort out the ambiguous details of complex decision situations and

> block actions that are irrational in a narrow time frame, but are rational when more distant consequences are considered. Emotions allow us to act in ways compatible with our long-term interests…. Many rational organizational strategies are pursued on highly emotional grounds, and much of what we describe as rational is in fact emotional.[12]

Fortunately, SmartDrill faced a situation in which rational processes were possible, and used those processes in an optimal way. The situation was not too complex to allow rational processes to be used, and the organization, after a great deal of hard work, was able to obtain and effectively interpret a wide range of relevant information. Equally fortunately, they refused to employ commonly used shortcuts when they could do otherwise. They refused to take rhetoric about the overriding cost advantages of the *maquiladora* workforce for granted. Instead, they carefully sought out more precise and more complete information than they were provided in the consultants' report, considered both the short- and long-term effects of their decision, and were willing to change their view of the situation when the facts of the case warranted their doing so. As we will point out in Chapter 8, the latter characteristic is unfortunately rare. They carefully examined the assumptions underlying the consulting group's recommendations, and sought out information that confirmed some of those assumptions, rejected others, and modified yet others. In the process they refined their understanding of the tasks that were being performed in their various operations and the type and level of skills necessary to successfully complete each of those tasks.

Emotional aspects of the decision-making process helped them focus their attention and to think about the long term, rather than reactively focusing on short-term cost savings from lowered wage rates. In sum, their decision-making process led to a more precise understanding of their own operations and of the *maquiladora* industry.

In an important way, SmartDrill's decision making was a harbinger of changes in the entire industry. Largely because of the recession of 2000–2002 in the United States, to which 90% of *maquiladora* production is shipped, the industry experienced its own recession. In addition, companies that had moved to Mexico solely in order to exploit its US$1.50 a day wage rates moved their operations to even lower-wage-rate countries in Asia and Central America. A quarter of a million jobs disappeared, and more than five hundred *maquiladoras* closed their doors. Eventually, the U.S. economy started to recover, and so did the *maquiladoras.* Estimates of the number of jobs added during 2004 range from 55,000 to 200,000, but even the most conservative estimates show that a recovery has begun.[13]

However, the industry is changing. The recession led many companies to reassess their strategic decisions, and to take the complex perspective that SmartDrill had taken. They recognize that Chinese and Central American wage rates do not compensate for the logistical problems of supplying a U.S. market from such a distance or through a relatively undeveloped transportation system. This is especially true for bulky products, because of their greater transportation costs, but it also is true of companies that need high levels of production flexibility in order to deal with rapidly changing market demands. Diane Velasco of the *Albuquerque* (New Mexico) *Journal* notes:

> Tech products often are outdated in a matter of months, compelling producers to keep manufacturing lead times as short as possible. [In addition] U.S. companies that require small runs, rapid turnaround or just-in-time delivery are also likely to choose Mexico over China, despite Chinese wages that are a fraction of the ... average paid in Mexican maquiladoras.[14]

High-tech companies also are reticent about moving to China because of its weak protection of intellectual property rights. Companies interested in the growing markets of Latin America also have good reasons to remain in Mexico. New trade alliances between Mexico and Asian countries have created even more complex decision situations. For example, in an operational system very much like SmartDrill's, Motorola's Chinese-run plant in Ciudad Juarez, Mexico, produces high-valued-added components that are combined with simpler components produced in China to produce products for the North American and European markets. As a result, companies interested only in low wages—cotton textile manufacturers, for example—are moving to or staying in Asia or Central America. In contrast, companies like SmartDrill are staying in or moving to Mexico for more precise reasons based on more complex decision making.

Like SmartDrill, they also are learning much more about the Mexican workforce, particularly in terms of level of education. Approximately 1 million workers enter the Mexican labor force every year, but most lack the training necessary for the increasingly sophisticated design, manufacturing, and assembly processes needed for the kinds of products that organizations are planning to produce in Mexico. The companies are becoming more and more choosy about who they hire—in terms of both education level and long-term residence in the local community—but also more willing to provide advanced training for a workforce that is "very trainable and willing to learn."[15] However, when companies move toward higher-value-added products, and increase the skill and educational level of their workforces, wage rates tend to rise, putting additional pressure on firm profits, which makes careful decision making even more important. SmartDrill took all of these considerations into account, and eventually made the kind of decision that other companies now are making and will need to make in the future.

Review and Study Questions

1. Use the global engineering model (GEM) to classify the facts of the SmartDrill case.

2. Describe how facts in the different layers of GEM interact to impact a decision that would only consider low wages as the main decision criterion.

3. Assume you must decide whether to close the plant in Houston and transfer all operations to Mexico. What would be your ranking of each option if you apply the AHP method? Compare your solution to the outcome of the case study.

4. What factors affected the decision maker's ability to make an ideal rational decision in this case study?

Notes

1. See Geert Hofstede, 1980b, 2001, especially chapter 8. Stephen Fineman discusses the ideological (and thus cultural) dimensions of theories of organizational rationality/emotionality in "Emotions and Organizations," in *Handbook of Organization Studies*, ed. Stewart Clegg, Cynthia Hardy, and Walter Nord (London: Sage, 1996), pp. 543–564, and in "Organizing and Emotion: Towards a Social Construction," in *Towards a New Theory of Organizations*, ed. M. Parker and John Hassard (London: Routlege, 1994).

2. Karl Weick and Larry Browning, "Argument and Narration in Organizational Communication," *Yearly Review of Management of the Journal of Management* 12 (1986): 243–259. For a comparison of individual and organizational decision making, see C. R. Schwenk and M. A. Lyles, "Top Management, Strategy, and Organizational Knowledge Structures," *Journal of Management Studies* 29 (1992): 155–174.

3. Markus Vodosek and Kathleen Sutcliffe, "Overemphasis on Analysis," in *Pressing Problems in Modern Organizations (That Keep Us Up at Night)*, ed. Robert Quinn et al. (New York: AMACOM, 2000).

4. The chess example is from A. Newell and Herbert Simon, *Human Problem Solving* (Englewood Cliffs, NJ: Prentice Hall, 1972). Memory and managerial decision making are examined by Kathleen Sutcliffe in "Commentary on Strategic Sensemaking," in *Advances in Strategic Management*, ed. J. Walsh and A. Huff (Greenwich, CT: JAI, 1997). Also see Henry Mintzberg, *The Rise and Fall of Strategic Planning* (New York: Free Press, 1994); and Vodosek and Sutcliffe, "Overemphasis on Analysis."

5. Karl Weick, *Sensemaking in Organizations* (Thousand Oaks, CA: Sage, 1995), p. 60.

6. Charles Lindblom, "The Science of Muddling Through," *Public Administration Review* 19 (1959): 412–21. Henry Mintzberg and Alexandra McHugh, "Strategy Formation in an Adhocracy," *Administrative Science Quarterly* 30 (1985): 160–97, provide an excellent example of a successful "muddling through" organization.

7. Richard Butler, Graham Astley, David Hickson, Geoffrey Mallory, and David Wilson, "Strategic Decision Making in Organizations," *International Studies of Management and Organization* 23 (1980): 234–49. The example is based on George Farris, "Groups and the Informal Organization," in *Groups at Work*, ed. Roy Payne and Cary Cooper (New York: John Wiley, 1981).

8. Leslie Sklair, *Assembling for Development: The Maquila Industry in Mexico and the United States* (London: Unwin Hyman, 1989).

9. Mariah de Forest, "Thinking of a Plant in Mexico?" *Academy of Management Executive* 8 (1984): 33–40.

10. Jorge Martinez-Vazquez and Duanjie Chen, "The Impact of NAFTA and Options for Tax Reform in Mexico," Policy Research Working Paper 2669 (World Bank: September 2001).

11. De Forest, "Thinking of a Plant in Mexico?"

12. Fineman, "Emotions and Organizations." Also see Richard de Sousa, *The Rationality of Emotion* (Cambridge, MA: MIT Press, 1987); T. D. Kemper, "Reasons in Emotions or Emotions in Reason," *Rationality and Society* 5 (1993): 275–82; and Dennis Mumby and Linda Putnam, "The Politics of Emotion," *Academy of Management Review* 17 (1992): 465–86.

13. J. Canas and R. Coronado, "Maquiladora Industry: Past, Present, and Future," *El Paso Business Frontier*, no. 2 (Federal Reserve Bank of Dallas, 2002); and Intelligence Research Limited, (2004), www.latinnews.com/ Maquiladera Industry: Past, Present, and Future. The latter source points out that China's wage advantage was increased by its acceptance into the World Trade Organization.

14. "New Mexico Development Board Brings Back Campaign to Lure Midwestern Factories," *Albuquerque Journal Online*, September 2, 2004, p. 3.

15. See ProLogis Research Group, "Is China's Economic Success a Threat to Mexico?" http://www.newscom.com/dgi-bin/prnh/19990420/PROLOGIS.

8

Do What You're Told, and Don't Confuse Me with Facts

Headquartered in the United States, American Automobile Components (AAC) is one of the world's leading automotive suppliers with approximately US$10 billion revenue in 2001. The company supplies components for more than forty vehicle manufacturers. AAC has more than 150 facilities operating in more than twenty countries on almost every continent. The firm's products are divided into three main categories: chassis systems (such as braking systems, steering systems, etc.), occupant safety systems (such as air bags, safety belts, safety electronics, etc.), and other automotive systems (engine components, body control systems, etc.). The chassis systems group brings AAC the largest revenue, with 60% of total revenue, while the remaining revenue is distributed to safety systems (30%) and other automotive groups (10%). The company employs more than 60,000 workers worldwide. For some time AAC has had five plants in various parts of Mexico. It recently set up a new *maquiladora* in Azteca, Mexico, near the U.S. border. Our research team was invited to visit this plant in order to consult with its managers in solving some engineering-related problems. The team's experience at the Azteca plant is the focus of this chapter and Chapter 11.

Cultures, Decision Making, and Organizational Learning

In Chapter 7 we explained that a core value of Western cultures, and especially of U.S. culture, is the belief that both personal and organizational decision making should take place through the application of a rational actor model. In that chapter we explained that there is a wide range of organizational decision situations in which rational decision making is impossible or unwise. Most organizational decisions are so simple and so repetitive that it makes little sense to treat each one as a separate event. Instead, organizations should create clear and unambiguous criteria for determining that there are existing policies for dealing with a situation, and devise standard operating procedures for dealing with those routine events. In this way, the organization spends its time and energy on nonroutine decisions—too complex or too rare to be managed through standard routines but simple enough to allow rational decision processes to be used.

Still other situations are too complex for strictly rational processes. However, Western/U.S. cultures require decision makers to at least *appear* to be rational, both for their own self-esteem and for their influence in the eyes of other people. In a provocative article aptly entitled "The Technology of Foolishness," James March explained that Western societies in general, and U.S. culture in particular, embrace three primary articles of faith:

1. *The Preexistence of Purpose*: People begin with goals, make choices based on these goals, and can offer adequate explanations of their actions only in terms of their goals.
2. *The Necessity of Consistency*: People choose to act in ways that are consistent with their beliefs and with their roles in their social groups (families, organizations, communities, and so on).
3. *The Primacy of Rationality*: People make decisions by carefully projecting the probable effects of different courses of action, *not by intuition* (in which they act without fully understanding why they do what they do) or by tradition or faith (in which they do things because they always have been done that way).[1]

A major part of acculturation in these societies involves learning these three commandments. In Western societies, people learn that children act impulsively, irrationally, and playfully. Adults act calmly and rationally, making decisions by carefully considering a number of complicated factors. They are spontaneous only when they have calmly and rationally decided to be spontaneous. Because people are products of their societies, their individual identities and self-esteem are linked to the belief that they are rational people. As a result, even when employees do behave in ways that are not strictly rational, they need to pretend that they have not. In fact, the drive to appear to be rational decision makers may be strongest when people make nonrational decisions. Doing so allows them to maintain an image of being calm, rational adults and to gain comfort from the knowledge that they are capable adults living in a stable, predictable, rational world. U.S. psychologist Edward Stewart concludes, "North American decision-makers do not observe rational decision-making in their own work and lives, as a general rule, but they restructure past events according to a decision-making model…. Thus in the United States rational decision-making is a myth."[2] But, social myths are important because they help people make sense out of their experiences, support their self-images, and legitimize their actions.

After all, organizational decision makers are people, just like everyone else. They sometimes base decisions on what they wish was true rather than what is true, succumb to groupthink instead of standing up for what they know is correct, and persist in failing policies long after the available evidence makes it clear that those courses of action should be abandoned. Like everyone else, organizational decision makers often search more actively for relevant information *after* they make decisions than before they do so—that is, they spend a great deal of time and effort *rationalizing* decisions they already have made.

Seventy-five years of research on the ways in which executives make decisions indicates that they are as likely to use information to rationalize their decisions as they are to use it to make them. Even when they do seek out information before making decisions, they may use a given item of information more because it is readily available than because it is accurate or relevant. It takes time and effort to obtain reliable information. In addition, seeking information usually involves admitting one's ignorance. In organizations in which *appearing* to be informed is rewarded and *appearing* to be uninformed is punished, it may be wiser to rely on information that is easily accessible than to search for better information that cannot be quietly found. In this sense, organizations are political creations more than they are strategic machines.[3] In normal organizational decision situations, employees make choices and then begin to construct, share, and publicize seemingly rational explanations and rationalizations of their choices.[4]

The rational actor model also presumes that the purpose of decision making is to solve problems. This presumption usually is accurate, but only to a point. It often is more accurate to view decision making as a political ritual in which employees pursue their individual agendas as well as the goals of their organizations. Sometimes these individual agendas are simple and tangible. One decision maker may support a building plan because it will give his subordinates more overtime. A department head may support the same decision because it will give her an opportunity to transfer two troublesome workers to another division. Other employees may agree with a proposal because it will divert upper management's attention away from the large equipment purchases that they plan to make during the coming weeks. When combined, the different individuals may use the decision episode a little like a garbage can, dumping into the discussion a plethora of concerns, only some of which are logically related to the problem being discussed. Of course, it is not likely that the participants will admit their real motives in public. Instead, they will search for a rationale for the building project that is acceptable to everyone and that can be stated in public. In this way, communication may be used to obscure the participants' real motivations as much as it is used to reveal them. In the process the participants make what seems to everyone to be a rational decision.[5]

Political considerations and power relationships may influence the decision-making process far more than the goal of making the best decision. Powerful employees can push an issue through the process rapidly or can slow the process down by pressuring for a longer information search, demanding that other interested parties be involved in the process, tabling the issue, or referring it to a subcommittee. For example, one of us once observed a decision-making episode involving the president of a subsidiary of a large multinational corporation, the chairs, and ten other people, including all of the organization's vice presidents and department heads. The group had to decide between the terms of an existing contract and a new pricing system. After a half hour it was clear that the president and executive vice president disagreed on the proposal. One senior vice president added information about international market conditions, but no other members spoke up because they realized that doing so could alienate one of the two top-ranking people in the organization. No action was taken, but another meeting was scheduled to discuss the issue further (and then another, and another, etc.).[6] In many organizations, employees attend meeting after meeting, year after year, where the same issues are discussed and the same arguments and information are presented. This repetitiveness is irritating, primarily to employees who believe Western cultural myths that rational problem solving involves solving problems—once and for all. For employees who realize that the purpose of meeting is meeting (that is, to enact a ritual), repetitive problem solving is easy to understand.

Plans serve as *symbols*, signals to outsiders that the organization really does know what it is doing. They also are *advertisements*, tools with which to attract investors or mobilize workers. Plans also are *games*, means of determining how serious people are about their ideas. Planning takes time and energy. Unless a person or group really is committed to the idea, they will not expend the effort needed to plan. The 3M Company is famous for cutting off the funding for its new projects multiple times, in order to cut their advocates back to the real fanatics.[7] Finally, as Chapter 7 pointed out, plans often are *excuses for further planning*. Because many decisions are too complex to be sorted out completely in a single decision-making episode, they must be managed incrementally. Decision makers muddle through complex problems by making a series of small decisions. Eventually, the minor actions that they take provide new information, and help them make sense out of complicated problems. In the process they act, and by doing so convince themselves, and others, that they really do understand those problems.[8]

In summarizing research on nonrational aspects of organizational decision making, we have not intended to disparage employees or question their competence. Instead, we believe that it is the rational actor myth that should be examined critically. Employees face complex decision situations, in highly politicized situations, armed with limited time, energy, and information. In spite of all of these limitations, they typically make decisions that are good enough to keep their organizations operating at acceptable levels of efficiency. The rational actor model ignores these complexities, and thus imposes an unrealistic set of assumptions on organizational decision making. In fact, employees usually compensate for weaknesses in the design and structure of their organizations so effectively that those organizations operate more efficiently than they should. But, one of the most important skills that engineers need to develop is to be able to detect nonrational processes in operation, and to minimize them whenever possible.

The Facts of the Case

AAC's Mexican operations have long had a problem with high inventory costs. In the past most of its plants experienced inconsistent inventory management: some parts were overstocked while there were shortages of other components in the same product family. Having too few parts can be a disaster because it forces a whole automobile assembly line to stop. Plant managers avoided the latter problem by increasing the holding level to make sure that materials were sufficient to support production at all times. Of course, this solution increases inventory-related costs. To complicate the situation further, each plant controlled its own inventory, has different production planning systems, and schedules its own distribution. As a result, one plant may have excess inventory of one component, while another plant in the system has too few items of the same component on hand to meet its obligations.

In order to deal with these problems, AAC-Mexico assigned Juan to head a project to improve the inventory management system and procedures. Juan had been working at AAC headquarters in Mexico for 7 years, and has a degree in industrial engineering from a university in Mexico City. Prior to working for AAC, he had worked as a purchaser for a large American automaker for 5 years. At AAC, Juan's initial responsibilities were to oversee purchasing for all AAC operations in Mexico. Soon he was promoted to head of the logistics and supply chain division. Currently, he manages raw material movement of all Mexico-based subsidiaries (plant to plant, supplier to plant, and plant to customer). In short, Juan has advanced through the company because of his ability to manage inventory, his supervisors view him as their resident inventory expert, and his annual evaluations, raises, and promotion potential all depend on his continuing to play that role well.

Soon after his most recent promotion, Juan decided that the key problem AAC faced was its decentralized inventory control system. He invited two members of our research team to serve as consultants to help him initiate a system of consolidating inventory control in two locations, AAC-Roca, located in the U.S. near the Mexican border, and AAC-Azteca, located in a Mexican border state. In addition, the Azteca plant would be a test site for all plans to reduce the company's inventory costs. If new ideas were successfully implemented at Azteca, they would be launched as a company-wide strategy. Juan also proposed that the company consolidate inventory in Border City, Texas. AAC already had a facility there, so using Border City would save the money that would be required to build a new warehouse. Juan was confident that this would decrease inventory levels significantly.

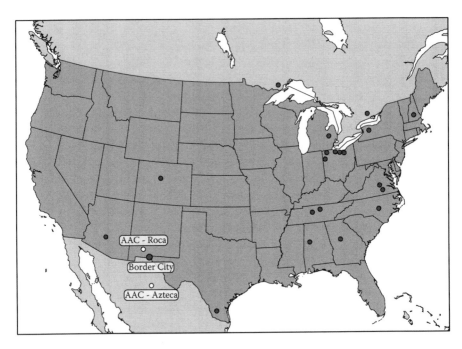

FIGURE 8.1
Location of Border City, and AAC plants in Mexico and the United States.

Juan decided to initiate a pilot project at Azteca focusing on TMG008, the code name for an electronic engine control assembly. The TMG008 components are supplied from several companies. Most of them are companies located in the U.S. Midwest and Southwest (see Figure 8.1), but one of the key suppliers is the AAC-Roca plant just north of the Mexico-U.S. border. Before the pilot project was implemented, the components and raw materials for the electronic engine control assembly were individually shipped by each supplier to the AAC-Azteca plant, where they are partially assembled. Azteca then shipped the partially assembled parts from northern Mexico to AAC-Roca for final assembly. Finally, AAC-Roca shipped the electronic engine control assembly as a finished component to automotive company customers in the United States.

TMG008 Demand and Inventory Analysis

Every week, AAC-Azteca receives an electronic file containing the actual demand for TMG008 assemblies to be shipped within the following week. The demand data for AAC's three customers are provided in the appendix. After looking at the initial demand data, Juan concluded that the demand for TMG008 assemblies was fairly stable since it is contract-based purchasing, there are only three customers, and their individual demand fluctuated little

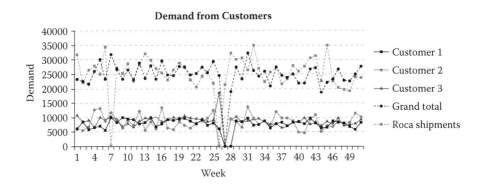

FIGURE 8.2
TMG008 weekly demand in 1 year.

during the year. Figure 8.2 shows TMG008 demand statistics for each customer, the combined customer demand, and number of partially assembled units that Azteca ships out to Roca every week for an entire year.

Initial Ordering Policy

Before the pilot program started, Azteca's ordering policy relied mostly on its planner's experience. According to Figure 8.2, the average Roca shipment to Azteca is approximately 25,000 units per week. AAC-Azteca reviews and generates orders weekly, with a lead time of 1.5 weeks from the time the order is placed to when it is actually received. The previous policy is illustrated in Figure 8.3. Juan realized that the ordering policy appears to be incurring high ordering costs. Juan immediately implemented a revised ordering policy.

Under the new, consolidated system, AAC-Border City served as a hub for both the Azteca and Roca plants. Suppliers sent materials to Border City, where it was warehoused until needed by one of the plants. Initially, the implementation of a centralized consolidation system and a simple

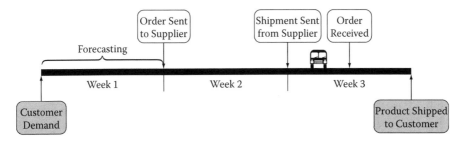

FIGURE 8.3
AAC-Azteca supply chain/ordering policy.

TABLE 8.1

Costs of Inventory versus Transportation Costs after EOQ
Implementation

Inventory on Hand	Holding Rate	Holding Cost/Year
$320,777.93	20%	$64,155.59

Cost of Materials	Transportation Cost	Transportation Cost/Year
$22,630,483.51	3.50%	$792,066.92

economic order quantity (EOQ) policy worked fine. Suppliers would ship
their items to Border City, and AAC took care of processing the documents
needed to bring the materials across the border. Inventory costs declined,
as Juan expected, but AAC's total budget savings were much smaller than
he had forecast.

Dealing with Transportation Costs

The consulting team pointed out that although inventory and holding costs
had been reduced in the new system, transportation costs increased because
each individual supplier now uses its own transportation system to ship its
items to Border City. Since total transportation costs are almost twelve times
as large as inventory costs, very small increases in them can easily wipe out
the savings gained from reduced inventory costs. Table 8.1 shows the inventory on hand costs calculated from the data in the appendix. The inventory
cost is estimated as 20% of the value of the material on hand, and the transportation cost is estimated around 3.5% of the cost of materials being transported. Given the annual demand of TMG008 of 1,291,200 pieces, the total
cost of material is $23.254 million, or $813,907 annually for TMG008 alone.
In addition, the centralized system requires extensive cooperative planning
among the various parties, which creates an increasingly complex operation
and scheduling system. Breakdowns in communication and coordination
become more likely, and the reliability that suppliers and customers have
become accustomed to is more difficult to sustain.

Juan decided that the best way to address the transportation cost problem
was to rearrange the raw material pickup routes and outsource transportation service to Fast Trucks Co., an American truck company. He clustered
all of his U.S. suppliers into three groups according to their geographical
location. Eight suppliers represented more than 98% of the total material
cost. Table 8.2 displays the distances of the six suppliers that are located far
from the assembly plants. It also shows the coordinates of the suppliers and
the capacity utilization of their transportation system. Weights reflect the
weekly volume transferred from each supplier. (The other two major suppliers are located so close to Border City that they will continue to ship their
materials directly.)

TABLE 8.2

Suppliers' Coordinates and Weights

Pi	Supplier	City	State	Latitude	Longitude	Weight
P1	Supplier1	Duncan	IN	39.20139	−85.921390	1.88
P2	Supplier2	Orangeville	SC	34.85250	−82.394166	1.00
P3	Supplier3	Eugene	IL	42.03722	−88.281111	0.12
P4	Supplier4	Whitestone	VA	37.08028	−77.997500	0.12
P5	Supplier5	Moris	OH	41.13833	−81.863888	0.12
P6	Supplier6	Dillsville	OH	39.75889	−84.191666	1.31
	AAC-Border City	Border City	TX	31.75861	−106.48639	

The suppliers who are located close to one another consolidate their shipments at a local warehouse, and then ship them to Border City together. Figure 8.4 illustrates the three primary routes: Rockland-Anderson-Munson-Border City, Whitestone-Chase-Border City, and Grand Prairie-Clark-Green Valley-Border City.

In an effort to coordinate transportation better, Juan and his team generated a weekly pickup and delivery plan based on the demand of each individual material and the geographic location of the supplier. The weekly plan is broken into individual reports and sent to Fast Trucks Co. and all suppliers.

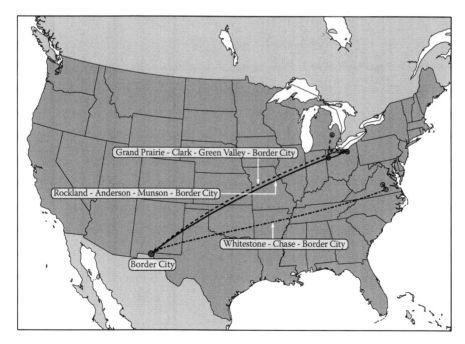

FIGURE 8.4

Three example routes.

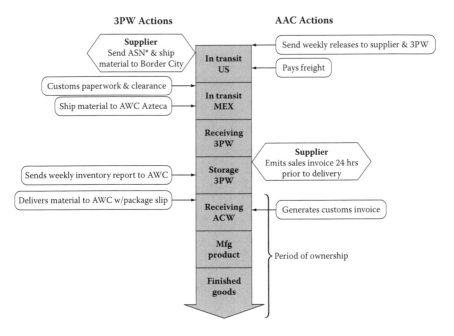

FIGURE 8.5
Third-party warehousing (3PW) contract activities.

Back to Inventory Control

The new transportation strategy seemed to work well, so well that it could exceed the warehouse capacity in Border City. Because the economy was sluggish, Juan did not want to invest in building a new warehouse or expanding the existing one. As an interim measure, Juan developed a new warehousing strategy called third-party warehousing (3PW). Outsourcing warehousing to a nearby firm provided not only material storage service for AAC-Mexico, but also shifted the burden of dealing with customs and paperwork when materials cross the U.S.-Mexico border and delivery to all manufacturing plants in Mexico. Juan anticipated that 3PW would help the company with the space problem, and return more productivity, less overall cost, better service, and faster response. The 3PW activities are illustrated in Figure 8.5.

A Year Later ...

When Juan implemented the new transportation system he decided to revisit the decision after 1 year to assess its effectiveness. He called a meeting with his operations manager, Tomas, and his purchasing manager, Miguel. They prepared their reports and met with him one afternoon.

Tomas began, "I don't have very good news for you, Juan. At first I thought that our new transportation plan was going to do great, but I have not seen any improvements in the costs. After looking at the numbers more closely, I can tell you what it is ... its utilization. Our average fleet utilization is at 59%, which costs us over $150,000 a month in transportation. The truck industry average utilization for cross-border transportation is over 80%. We are paying a lot of money to move air!" Moreover, "even if we could increase fleet utilization to 90%," Tomas continued, "the cost would only go down to $100,000. That's not very much. We need to think of another approach, and soon. I don't think AAC headquarters is going to like the numbers at all."

Juan asked, "Do you have any ideas, Miguel?"

Miguel's department had been in charge of keeping the supplier routes up to date with weekly purchasing requests. He responded, "Well yes, Juan, I have been keeping in contact with Fast Truck Co. and our suppliers. There is one thing I don't understand, and that is why we have to hire an American truck company to do our transportation. We are already using a Mexican truck company to carry our items from Border City to each individual plant. Wouldn't it be easier and cheaper if we started using the same Mexican truck company to do it all?"

"Actually, it's not that simple," said Juan.

"How come?" asked Tomas. "Isn't Mexico in NAFTA now? What about the Free Trade Agreement? I heard that Canadian trucking companies can come in and drop off stuff anywhere in the U.S. And an American driver can go across to Canada too. I thought we were included in that agreement."

"Okay. If I remember correctly, NAFTA went into effect in 1994, right? Here is the deal. The provision of free transportation should have started in 1996. That would mean that all truck drivers from Mexico, Canada, and the U.S. could enter anywhere they want within those countries' boundaries. However, this provision brought up heated debates among truck driver unions/trucking associations in the U.S. They were worried that Mexican drivers could steal away their jobs due to the great differences in wage. Then, NAFTA decided not to go on with this provision but allowed drivers from these two countries to drive up to 20 miles from the boundary, no more, no less."

"So what is going on now?" asked Tomas.

"This NAFTA decision has had a great impact on bilateral trade. Let's say you want to move your goods from Mexico City through New York. First, the Mexican truck company can take the items up to any port of entry along the border, mostly in Texas. Then they have to unload everything at the crossing docks there. A third party called a drayage company crosses the load from Mexico to the U.S. Close to the border, the load is reloaded again on American trucks. Then your goods can go off to New York. This simple movement requires tons of documents and involves several different parties, like both the Mexican and American truck companies, brokers, and government agencies, because most crossing docks are controlled by the U.S. government. This results in higher costs, longer lead times, delays, etc. And that is what we have to go through to transport materials from our suppliers," said Juan.

"It gets very complicated!" said Miguel.

Juan continued, "One more thing, if you think that it would be cheaper to have Mexican truck companies, believe it or not, Mexican transportation cost is normally higher than American."

"Are you serious? Didn't you just say the wages of Mexican drivers are cheaper?" asked Miguel.

"Yes. However, most Mexican trucking companies don't have big trucks. So, it's hard to get the scale economy advantage in the long term. Also, due to the fact that road conditions in Mexico are poorer, the company has to pay more in maintenance to keep the trucks in good shape. Things like that make transportation costs higher in Mexico.[9]

"Well, Miguel and Tomas, thank you for your reports. I think we have a lot of work to do. Why don't we meet again next week, after we all have a look at the reports, and see if you come up with any ideas on how we can fix this," Juan finished.

After the meeting Juan was worried. AAC's upper management had given him a mandate to lower inventory costs to half of last year's levels. The decision to focus time and money on transportation costs was based on the recommendations of the consulting team, not on commands from above. He knew that senior management had invested a lot of money to upgrade the Border City facility and put resources into his programs. Since those programs had been his idea, not theirs, he knew he was under a great deal of pressure to show major financial gains from them. And, he knew that he may not have any more chances to recover his performance.

Juan Revises His Plans

Juan sat in his office, looking at the information regarding the transportation situation. He had divided the thirteen groups of suppliers by their locations. Juan thought to himself as he looked at the map where all the suppliers are located, "Most of our suppliers seem to be located in the same area." He now knew that he had to think in terms of capacity utilization (not "moving air") as well as in terms of an efficient route structure.

The thirteen routes used to pick up raw materials show that most suppliers are from the Midwest. For the three routes mentioned earlier, Juan received an initial estimate of the costs per route and the utilization for each (Table 8.3). These numbers indicated to Juan that sooner or later, he had to rearrange the routes.

When he first designed the route structure he had not taken into account each supplier's inventory trends or its individual consumption of space in the trucks. Since Fast Truck's contract paid them by distance, they had no incentive to make certain the trucks were full. In fact, "moving air" reduces fuel costs, so Fast Truck's profits were higher if capacity utilization was low.

TABLE 8.3

Breakdown of Weekly Cost of Current Routes

City	Weekly Demand (Pallets)	Route	Numbers of Pallets	Utilization	Cost per Route (to Border City)	Cost per Pallet
Rockland	25	A	36	69.23%	$2,238.00	$62.17
Anderson	8	A				
Munson	3	A				
Whitestone	4	B	12	23.08%	$2,200.00	$183.33
Chase	8	B				
Grand Prairie	11	C	19	36.54%	$1,926.26	$101.83
Clark	6	C				
Green Valley	2	C				
Total cost per week					$6,364.26	

To make matters worse, the pickup schedule was not flexible enough to allow for the trucks to wait in the Midwest until they were closer to capacity. Therefore, Juan thought, "There must be a way I can rearrange these routes to be more efficient."

Managing Inventory

One week after Juan had originally met with Miguel and Tomas to discuss the problems with transportation, they met again to brainstorm possible solutions. Juan had an idea he wanted to present to the others, but there were a lot of complications that could make it nearly impossible to implement.

"Good afternoon, Miguel, Tomas," began Juan. "I wanted to begin with an idea I have been considering, and maybe you can give me some feedback, and thoughts as to how this can be implemented. If most of the suppliers are close together, we may have to move our consolidation point to somewhere that is closer to the suppliers, and consolidate our inventory from there. However, with the consolidation point being so far away, this could cause problems as to how we manage that inventory, I know we are already having problems as it is ..." At the time, Juan and his team were the ones that managed the inventory level at the consolidation point. Increasing the distance between Border City and the consolidation point would make it more difficult for him to monitor inventory, and would make AAC more dependent on information from the companies to which he had outsourced those functions. Juan knew that some of the information he received from production and planning departments was inaccurate and outdated. He realized that his new consolidation proposal would leave him even more dependent on effective sharing of information across the various group players.

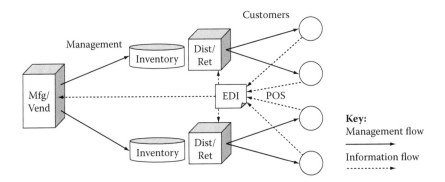

FIGURE 8.6
Vendor-managed inventory diagram.

Miguel, who as purchasing manager was especially sensitive to these issues, responded, "Move farther away from Border City? How are we supposed to keep track of everything with the warehouse being so far away, and who will be in charge? We need to think about this ..."

Tomas thought for a minute, and then broke in, "You know, I have heard of this new popular vendor-managed inventory (VMI) system. It's a distribution system in which inventory at the distributor is managed by the manufacturer. This is supposed to give the distributor, in this case us, a clearer picture of what the actual customer demand is."

"Really? This sounds interesting; this may be just the thing we need to make my plan work, can you explain more about it?" asked Juan.

"Sure, I can tell you all I know. I read about it in one of the trade magazines we receive. Let me draw a picture for you [Figure 8.6]. Basically, the manufacturer calculates the order quantities and manages the product mixes and safety stock levels for the distributor. In turn, the distributor provides the manufacturer with demand information. As a result, the manufacturer [vendor] is responsible for replenishing the inventory of the distributor without the need of a purchase order. On the other hand, the manufacturer gets more accurate demand information, reducing the need for unnecessary safety inventory. The point being that we could, in the long run, create savings for all parties in the supply chain and meet the demand more effectively. The catch is that we need to invest in electronic data interchange (EDI) technology so that our distributors can provide us point of sales (POS) data instantly. Also, we need to develop excellent forecasting capabilities."

"So one of the major advantages of VMI is that given more accurate demand patterns, AAC can yield lower inventory investment," responded Juan.

"Yes, that is the point," said Tomas.

"Well, maybe you can look into that and tell me how that would work," responded Juan.

Complications with Sharing Information

Tomas and Juan met later that week to discuss the preliminary plans.

"Thanks, Tomas. I was just a little curious about how things are going," Juan asked.

Tomas replied, "Let me show you something. Without VMI, both the manufacturer and the distributor will plan for their own inventory levels and safety inventories at their respective facilities. As you see, both of them own and control their safety stocks. This is a redundancy and is causing excessive inventory and hidden costs. Now, with this new plan, one of those plants can ignore safety stock. Also, with the integrated information, the products will be ready to ship to our suppliers within 2 weeks, compared to the current situation, where product is shipped at the end of the third week. Moreover, with the additional coordination of both materials systems, the problem of transporting air should be minimized."

"Good job. Let's present this to headquarters so that we can begin to implement this project."

"The first phase will be the most challenging one. Once we can get through it, the second phase to all the plants in Mexico will be a lot easier," replied Tomas.

Juan continued to use AAC-Azteca and AAC-Roca, as pilot plants for VMI. In this case, the Azteca and Roca situation is quite complicated because they both play the role of supplier and buyer. Roca supplies raw materials to Azteca. Once materials have been processed, Azteca will ship intermediate products back to Roca for further operations, as a typical *maquiladora* plant. Therefore, when Roca has a problem and cannot deliver materials on time, Azteca will have problems too. Then, Azteca's problem will impact Roca's production schedule again. This creates some tensions and distrust between them.

When material movement managers of both sides learned that they were going to implement VMI, they showed reluctance to participate in this program. "If we let them control our inventory, it will be disastrous," "I never trust those guys," and "How could I level up the trust when I still have troubles dealing with them?" both managers said.

Having worked with the team, Juan came up with the idea that Azteca can prepare the production planning one week ahead during Tuesday to Thursday. This plan will be sent to Roca by Friday of Week 1 so they can start the production line on Week 2 as scheduled. Roca can determine the inventory level by subtracting the number produced from the current week and demand received for next week. In Figure 8.7, the current supply chain phases are illustrated, where it takes approximately 3 weeks to get products. Figure 8.8 illustrates the supply chain phases if VMI were to be implemented.

Even though implementing VMI required an additional investment in computer technology, AAC's upper management went along with Juan's proposal. Unfortunately for Juan, when time came for his annual review, AAC's management made it very clear that they were not impressed with his efforts

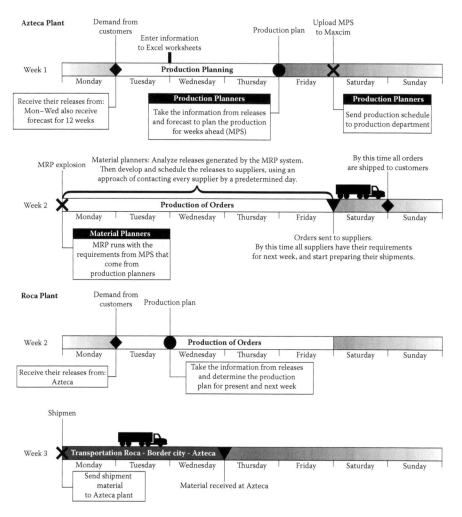

FIGURE 8.7
Supply chain phases without VMI.

to reduce transportation costs. They had hired him to deal with excess inventory costs, not to tinker with the entire system. They had gone along with his transport proposals because they promised to improve inventory control, not because they were persuaded that transportation costs were a bigger problem, in spite of what the consultants found. Even if his transportation system had worked perfectly, and it had not, management would not have been impressed unless it had reduced inventory costs. They congratulated him on the progress he had made in reducing inventory, and made it clear that his next annual review would focus on that one issue.

After the meeting Juan was even more worried. AAC's upper management was narrowly focused on achieving numerical goals. It was true that on paper

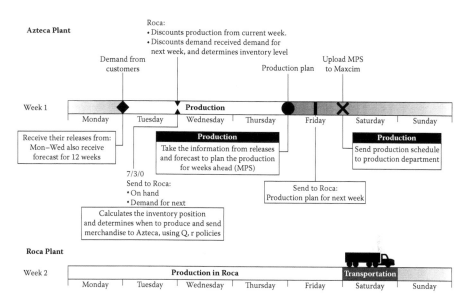

FIGURE 8.8
Supply chain phases with VMI.

there was a reduction in the size of the inventory kept on site in the various plants. But, those numbers were something of an illusion. AAC's inventory figures did not include material that is located in Fast Truck's trailers or the material stored in the 3PW warehouses. As a result, inventory control had been much more successful on paper than on the ground. But, Juan's bosses did not seem to care about the realities that he faced. He was being held accountable for only one thing—inventory—and on-paper numbers related to inventory control. The larger system, and the real-world efficiency of that system, did not seem to be important. Juan started to feel trapped in a web of numbers, and to resent it very much.

Interpretation

In Chapter 7 we explained that a distinctive element of U.S. culture is faith in a rational actor model of individual and organizational decision making. However, we also noted that some of the key assumptions of the model were unrealistic—it is only under limited circumstances that truly rational decision making is possible, or required. Decisions often are based on decision makers' taken-for-granted assumptions rather than on a careful, systematic analysis of the available evidence. The wisdom of SmartDrill's decision makers, especially Bob, was based on their willingness to critically examine their taken-for-granted assumptions. Everyone—SmartDrill's consultants, key suppliers in Mexico, and a massive *maquiladora* industry—*knew* that U.S. organizations should move production operations to Mexico in order to benefit from its lower wage rates. But, SmartDrill was not willing to take that assumption for granted, and by doing so came to a very different conclusion. By refusing to treat assumptions as *assumptions* they made rational decision processes work.

AAC's experience was very different. Its management started with the assumption that the primary problem it faced in its borderland operations was controlling inventory costs by reducing excess inventory. It hired Juan because of his expertise in that area, established inventory control as the primary basis on which his performance would be evaluated, and supported a series of decisions because they believed those steps would solve the inventory problem. Even when Juan demonstrated that transportation costs were a much greater problem, AAC's upper management interpreted those costs as part of the inventory problem, and funded Juan's proposals on that basis. So, from their perspective, it made perfect sense to evaluate Juan's performance narrowly, in terms of his ability to control inventory. From Juan's perspective, this focus made little sense because inventory was only part of the problem. He had learned from the Border City experience that trying to control inventory alone gained little for the company, and might even make the situation worse by driving up transportation costs.

However, three strains of research on decision making in U.S. firms help make sense out of management's position. The first was described briefly in Chapter 7. Once decision makers take a position on a key issue, it is very difficult to change directions midstream. This is partly for cognitive reasons: once people define a problem in a particular way, they seek out information to confirm their assumptions and tend to ignore or deemphasize disconfirming information. Political factors also are involved. Managers' power and influence is based on their decision-making skills, and admitting errors can threaten that power, especially in highly political environments. These tendencies are greatest when decisions are public. Once AAC's management defined the borderland problem as inventory control, formally included it in Juan's performance evaluation, and devoted a great deal of money to implementing his proposals, they were publicly committed to that definition. As management scholar Alan Tegar pointed out in the title of one of his most influential books, decision makers often have "too much invested [in a project] to quit"—too much emotional commitment, too much political capital, and too many tangible resources.

As a result, organizations sometimes create reward systems that either fail to achieve their goals or sometimes even undermine them. In what has become a classic essay on reward systems, Steven Kerr provided a number of examples of the folly of rewarding A while hoping for B. For example, in politics, U.S. citizens presumably want candidates for office to make their goals, proposals, and funding systems perfectly clear so that they can make informed choices. But repeatedly, voters reject candidates who do so and reward those who deal with images and personalities rather than issues and solutions. Citizens want state adoption agencies to place children in good homes, but enact regulations that base their administrators' budgets, prestige, and staff size on the number of children enrolled (that is, the number *not* placed). Consequently, they are encouraged to make it difficult to adopt these children. Universities are supposed to teach students, but reward research activities that have only an indirect positive effect on teaching quality. Students are supposed to go to college to learn something, but are rewarded by employers and graduate schools largely based on the grades they receive, regardless of what they have learned, thereby encouraging them to take easy classes (which reduces their opportunities to learn), obtain and study for passing exams in a course instead of for mastery of the material, and so on. Lower-level employees in manufacturing firms often perceive that they are rewarded for "apple polishing" and "not making waves" when upper management sincerely believes that they are encouraging all employees to be creative and innovative.[10] In AAC's case, the reward system that was established for Juan told him to focus his attention on inventory, and only on inventory, even if he knew that other factors were taking a much larger financial toll on the company.

The power and prestige afforded managers in U.S. firms also are based on their presumed expertise. While they may have obtained their initial jobs because of their technical expertise, their credibility *as managers* depends on their managerial expertise. This is why firms typically require new managers to enter MBA programs soon after they are promoted to managerial positions, and usually are willing to fund that part of their employees' education. In turn, managers demonstrate their expertise by learning and implementing the newest managerial techniques. In the best of circumstances, managers focus on the needs of their organizations, and implement new techniques carefully, and only after they have carefully investigated the extent to which those techniques meet the specific needs of their organizations. In the worst of circumstances, they implement new managerial techniques because they are new—a process of chasing managerial fads without a careful assessment of their fit with an organization's needs.[11] Although these fads may enhance a manager's prestige, they often do little to help their companies. For example, during the 1980s downsizing was the dominant fad in U.S. organizations. During the 1990s, the dominant fad was outsourcing. In fact, during both time frames, there was a stronger correlation between implementing the new strategy and managerial rewards than there was between firm performance and management's rewards.[12] Unfortunately, there soon was clear evidence that both techniques hurt organizations more than they helped, especially over the long term. Downsizing robbed organizations of much of their talent, eliminated hundreds of person-years of experience, and had a devastating effect on morale and motivation.[13] Outsourcing can significantly reduce an organization's costs, at least on paper, but as Chapter 5 pointed out, it significantly increases problems related to communication, coordination, and control. Juan sensed this problem when he realized that moving the central distribution point far from Border City could increase efficiency, but would reduce his ability to oversee and control the operation.

Of course, most organizational decisions involve complex mixtures of personal, political, rational, intuitive, and faddish dimensions. People are complicated beings, and their actions and decisions reflect that complexity. Juan's decision to accept Tomas's recommendation regarding the VMI system in some ways was another example of applying a managerial fad—a new technique that a decision maker had learned about that seemed to fit a situation in which everything else seemed to fail. Because it involved the application of new computer technologies, it was consistent with the dominant managerial fad of the early twenty-first century. During the past 50 years, organizational researchers have found that the application of new technologies often is more difficult, more expensive, and more unpredictable than their advocates imagined. Juan and his decision-making group probably did not investigate the impact of the VMI system to the extent that SmartDrill investigated opening a *maquiladora*, for example. In other ways, the VMI decision was an extension of AAC's management's obsession with inventory control. It focused the group's attention back on that single issue and on the relationship among the borderland plants. In the process, VMI brought the solution closer to home, where Juan could better monitor and control its effects. So, it simultaneously enacted some elements of rational decision making and some elements of less rational processes.

It is too early to tell if the VMI system will solve AAC's inventory control problems, or create the kind of improvement that Juan will need to show AAC's management at his next performance review. In theory, VMI provides just what AAC needs—a framework for synchronizing inventory and transportation decisions in ways that minimize distortion of information shared among the various parties and maximize efficiency. There are many success stories for the system, the most important one being the Wal-Mart Corporation. But, success stories abound during the initial stages of new fads, in part because no one wants to talk about the complications and the failures. For example, when multiple parties are involved in a VMI supply chain, each party has incentives to increase its own stock in order to not be caught short, creating a bias toward overstocking. In addition, vendors often have incentives to hold on to small orders until an optimal shipment size is reached. But, in other combinations of incentives, vendors are encouraged to understock materials. As a result of the competitive dynamics of multiple-party VMI systems, costs are substantially higher than VMI theory would predict.[14] But, the VMI decision, and all of the decisions leading up to it, provides a superb example of how organizational decision making often occurs, with all of its many complications.

Review and Study Questions

1. Classify the case study facts using the GEM model.
2. What difficulties do manufacturers face when sending materials to and from facilities on different sides of the Mexico-U.S. border?
3. Were there significant cultural factors affecting the success of Juan's project to reduce inventories?

Appendix

Weekly Demand from AAC's Three Customers and Roca Shipments

Week	Customer 1	Customer 2	Customer 3	Roca Shipments
1	6,056	6,368	10,740	31,644
2	8,474	5,504	8,531	22,003
3	5,904	6,836	8,856	26,456
4	6,649	12,605	6,665	27,950
5	7,030	13,103	9,914	24,874
6	5,630	8,803	8,896	34,323
7	9,950	11,706	10,125	0
8	8,261	9,196	9,169	27,226
9	9,993	6,273	6,894	25,153
10	9,543	8,949	7,837	28,501
11	9,317	7,225	6,439	22,509
12	7,740	12,148	8,843	28,141
13	8,269	5,620	9,627	31,879
14	9,960	8,545	9,375	29,935
15	6,759	6,148	10,293	26,934
16	7,770	13,396	8,404	25,568
17	9,121	6,422	9,166	23,079
18	8,878	5,779	9,836	26,313
19	9,575	8,483	9,624	28,815
20	9,786	7,391	10,316	27,692
21	8,669	6,189	9,809	22,987
22	7,665	8,035	9,548	20,534
23	9,327	8,846	9,147	24,128
24	7,209	9,817	8,388	25,626
25	8,005	12,296	9,043	21,878
26	5,996	0	18,423	0

Continued

Weekly Demand from AAC's Three Customers and Roca Shipments

Week	Customer 1	Customer 2	Customer 3	Roca Shipments
27	0	0	0	21,377
28	0	9,574	9,317	32,228
29	8,721	8,494	10,234	29,982
30	8,457	6,603	8,326	30,452
31	9,712	13,674	8,798	26,354
32	7,365	9,688	9,230	35,013
33	7,401	7,444	9,444	26,796
34	8,786	8,682	8,570	22,415
35	6,363	6,795	7,616	25,587
36	7,749	11,987	7,746	27,144
37	8,342	9,681	6,410	21,527
38	7,141	9,742	6,877	23,224
39	8,180	8,684	8,203	27,829
40	8,562	4,888	8,270	25,874
41	7,728	4,494	9,709	27,558
42	9,677	9,301	7,579	30,542
43	7,919	10,994	8,164	31,393
44	6,381	5,100	7,080	22,470
45	6,595	6,985	8,533	34,808
46	8,703	6,729	7,728	22,470
47	8,285	8,446	9,818	20,453
48	7,887	7,851	6,979	19,575
49	6,780	8,129	7,588	19,211
50	5,820	11,405	7,746	23,940
51	8,259	10,120	9,177	23,760

Notes

1. James March, "The Technology of Foolishness," in *Ambiguity and Choice in Organizations*, ed. James March and Johann Olson (Bergen: Universitetsforlaget, 1970). Also see Michael Cohen and James March, *Leadership and Ambiguity*, 2nd ed. (Boston: Harvard Business School Press, 1974). Harrison Trice and Janice Beyer take an even more direct position, arguing that rationality is *the* core assumption of organizations in Western societies, including the United States. (See *The Cultures of Work Organizations* [Englewood Cliffs, NJ: Prentice-Hall, 1993], especially chapter 2).

2. E. C. Stewart, "Culture and Decision-Making," in *Communication, Culture, and Organizational Processes*, ed. W. B. Gudykunst, L. P. Stewart, and S. Ting-Toomey (Beverly Hills, CA: Sage, 1985), pp. 177–211.

3. Alan Tegar, *Too Much Invested to Quit* (New York: Pergamon, 1980); M. Cohen, J. March, and J. Olson, "A Garbage-Can Model of Organizational Choice," *Administrative Science Quarterly* 17 (1972): 2.

4. Karl Weick, *The Social Psychology of Organizing*, 2nd ed. (Reading, MA: Addison-Wesley, 1979).
5. Cohen, March, and Olson, "A Garbage-Can Model"; Richard Butler, David Hickson, David Wilson, and R. Axelsson, "Organizational Power, Politicking and Paralysis," *Organizational and Administrative Sciences* 8 (1977): 44–59. For a revision of the original garbage can model, see Michael Masuch and Perry LaPotin, "Beyond Garbage Cans," *Administrative Science Quarterly* 34 (1989): 38–68.
6. Richard Butler, Graham Astley, David Hickson, Geoffrey Mallory, and David Wilson, "Strategic Decision Making in Organizations," *International Studies of Management and Organization* 23 (1980): 234–249. The example is based on George Farris, "Groups and the Informal Organization," in *Groups at Work*, ed. Roy Payne and Cary Cooper (New York: John Wiley, 1981): 95–120.
7. Thomas Peters and Nancy Austin, *A Passion for Excellence* (New York: Random House, 1985).
8. Charles Lindblom, "The Science of Muddling Through," *Public Administration Review* 19 (1959): 412–421. Henry Mintzberg and Alexandra McHugh, "Strategy Formation in an Adhocracy," *Administrative Science Quarterly* 30 (1985): 160–197, provide an excellent example of a successful "muddling through" organization.
9. In June 2004, the U.S. Supreme Court effectively ended the legal fight that limited Mexican trucks' access to U.S. roads. However, the legal changes had no effect on the factors that contributed to higher costs for Mexican firms. As a result, Bill Webb, president of the Texas Motor Transportation Association, concluded: "I honestly don't believe it's going to have a very big impact" (Patty Reinert and Jenalia Moreno, "Court Clears Path for Mexico's Trucks," *Houston Chronicle*, June 8, 2004, p. 1A).
10. Steve Kerr, "On the Folly of Rewarding A While Hoping for B," *Academy of Management Journal* 19 (1975): 769–783.
11. There is a rapidly growing literature on management fads, most of which is written by scholars outside of the United States. Among the best sources are E. Abrahamson, "Management Fashion," *Academy of Management Review* 21 (1996a), 254–285; T. Clark, "The Fashion of Management Fashion: A Surge Too Far?" *Organization* 11 (2004); J. W. Gibson and D. V. Tesone, "Management Fads: Emergence, Evolution, and Implications for Managers," *Academy of Management Review* 15 (2001), 122–33; B. Jackson, *Management Gurus and Management Fashions: A dramatistic Inquiry* (London: Routledge, 2001); and A. Kieser, "Rhetoric and Myth in Management Fashion," *Organization* 4 (2001), 49–74.
12. Alan Downs, *Corporate Executions* (New York: AMACOM, 1995).
13. M. Hitt et al., "Rightsizing: Building and Maintaining Strategic Leadership and Long-Term Competitiveness," *Organizational Dynamics* 23 (1994): 18–32.
14. Gerard Chachon, "Stock Wars," *Operations Research* 49 (2001): 658–674; Sila Cetinkaya and Chung-Yee Lee, "Stock Replenishment and Shipment Scheduling for Vendor-Managed Inventory Systems," *Management Science* 46 (2000): 217–232; Siddharth Mahaajan and Garrett van Ryzin, "Inventory Competition under Dynamic Consumer Choice," *Operations Research* 49 (2001): 646–657.

9

The Wisdom of Getting Everyone Involved: Communication and (Un)coordination at HOCH

HOCH (another pseudonym) is a German parts manufacturer for the automobile and truck industries. It has manufacturing and sales operations in the United States, Brazil, United Kingdom, South Africa, Japan, India, and Eastern Europe. Customers in Europe (71% of sales) and North America (20% of sales) include Ford, DaimlerChrysler, and Volkswagen, which purchases 20% of the company's total products worldwide. Recently, HOCH has tried to move into the Asian market, where its primary customers are Nissan-Renault and Toyota, but has found the process to be difficult and slow. Most of its design work is conducted in Europe, where 900 people are employed in research and development, and the United States (120 R&D engineers), although its Latin American and UK subsidiaries also design products for their local markets or adapt German designs to meet local needs (for example, it has 60 R&D personnel in Brazil and 15 in Mexico). Figure 9.1 depicts HOCH's locations worldwide.

HOCH has been operating in Mexico since 1975 and today is an important part of the Mexican automobile and truck manufacturing industry. The plant was built to provide drive train components for the Volkswagen Beetle (called a Sedan or Vocho in Mexico) that was built and sold in Mexico. Over time, HOCH-Mexico expanded its client list, and now dominates 85% of the Mexican market. In fact, it has only one serious competitor in Mexico, another German firm, which produces equipment for larger vehicles than those for which HOCH-Mexico serves, and is oriented toward markets outside of Mexico. The plant employs approximately 300 people, including 120 production workers. The general manager and two of the fourteen section managers are German expatriates. German engineers regularly visit HOCH-Mexico, and Mexican engineers often travel to Germany. The company also has an intern exchange program between Mexico and the home office, and two German interns currently are on site in Mexico. Because so few Mexican workers speak either German or English, no Mexican workers have interned in Germany. In 2000 HOCH-Mexico's sales exceeded US$51 million, a 14% increase over the previous year. Even during the Mexican recession of 2001, the company's sales increased 3%, and it expects to grow at a rate of 8% per year in the near future.

FIGURE 9.1
HOCH's locations worldwide.

Long Histories, Different Cultures

As we explained in previous chapters, Mexico has relatively high uncertainty avoidance (UA) scores. Germany's are much lower. However, when compared to other relatively wealthy countries, Germany's UA scores are quite high, more than twice those of Sweden and Denmark and almost twice as high as those in the UK. Hofstede concludes that this is characteristic of young democracies, which obtained their current political systems after WWI and as the result of losing a war. Both tend to create high levels of social and political anxiety. While Mexican culture encourages its citizens to manage uncertainty through complex webs of stable interpersonal relationships, Germans manage it through tight structures and scientific modes of thought. German organizational theories are highly rationalistic, and German managers focus much more attention on detailed planning and short-term feedback than in lower-UA cultures (for example, the United States or UK). Planning and monitoring provide security and stability. German managers expect forceful debate about organizational issues during decision-making episodes, even from their subordinates, and interpret a lack of active dissent as evidence of agreement.[1] Comparisons of German and U.S. engineers have had similar results. German engineers list anxiety avoidance as far more important to their professional life. They want to work in well-defined job situations, in which task requirements are made clear, want to be given detailed instructions about how to proceed with their jobs, and are more satisfied and productive when they feel little tension or stress.[2]

Since German national culture is more sensitive to uncertainty than their overall UA scores would indicate, it appears that the greatest difference in

Mexican and German national cultures involves power distance (PD). As earlier chapters have explained, Mexico is a very high PD culture, especially among its relatively uneducated workers; Germany's PD scores are among the lowest in the world, roughly half as high as those of France, and a little over one-third of Mexico's. In low-PD cultures, organizational structures tend to be flat, power is shared across all levels of the organizational hierarchy, decision making is decentralized, which leads to relatively small numbers of supervisors, and innovations occur only if they have a persuasive champion. Open displays of differences in status, power, and income are frowned upon, communication is open and direct, and information is expected to be readily available to any workers who need it.[3]

For Mexican workers, these organizational practices are likely to be disconcerting. Lines of authority and responsibility—a primary source of uncertainty in Mexican culture—will seem to be fragmented and ambiguous. German managers will solicit much more input into decision making than Mexican subordinates will feel comfortable providing, especially if they are not part of long-established work groups. The communication style of German managers will seem to be far too blunt and issues of face management will be quite common. Conversely, German managers are likely to view Mexican subordinates as unable to take initiative, excessively deferential and dependent, and even lacking in honesty and directness. However, the long-term relationship between HOCH-Germany and HOCH-Mexico should have helped employees in the two organizations to learn to work with one another and understand one another's communication styles, much more than a comparison of their national cultures would suggest. While intercultural problems do occur, they tend to be among employees new to HOCH-Mexico or between HOCH-Mexico and outside firms.

The Facts of the Case

HOCH-Mexico entered into a project with three other organizations to design and build a hybrid gasoline-electric engine for automobiles to be manufactured and sold in America. In the hybrid engine project, HOCH-Mexico must work most closely with Company B (a Japanese firm) because its contribution, the flywheel, must match up perfectly with the torsional damper. This type of flywheel is more complicated than those included in standard internal combustion engines. It turns with the crankshaft of the gasoline engine and stores kinetic energy. Thus, it contributes to the energy efficiency of the hybrid engine. The torsional damper is a combination of springs and washers that cushions vibrations of the crankshaft-flywheel assembly, reducing noise and eliminating unnecessary wear in the components surrounding it. Figure 9.2 shows a typical flywheel used in hybrid vehicles, Figure 9.3 shows a section view of the torsional damper, and Figure 9.4 depicts the location of these two elements in a hybrid engine assembly.

FIGURE 9.2
A typical flywheel. (Courtesy of HOCH.)

FIGURE 9.3
Section of a torsional damper. (Courtesy of HOCH.)

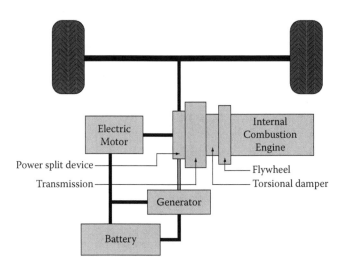

FIGURE 9.4
Assembly of a hybrid engine.

Hybrid vehicles use a small gasoline or direct injection diesel engine for light load operation and to charge an on-board battery. An electric motor then provides additional power for acceleration and higher loads. The hybrid approach can deliver extremely high fuel economy and ultra-low emissions without sacrificing driving range, convenience, or driving performance. It also eliminates the need for an electrical plug and charging station to recharge the on-board battery, and it allows the use of a much smaller battery to reduce weight, cost, and bulk. Power output from the gasoline engine is split between the drive wheels and a generator. The generator, in turn, is used to run the electric motor and to recharge the batteries. Operation of the hybrid combination is seamless and virtually imperceptible to the driver and passengers. Its five main operating modes are:

1. When pulling away from a stop or under a light load, only the electric motor powers the vehicle.
2. For normal driving, a combination of gasoline and electric power is used.
3. Under full-throttle acceleration, the electric motor receives additional power from the batteries.
4. During deceleration or braking, the electric motor functions as a generator to recharge the batteries.
5. The batteries are regulated to maintain a constant charge. When charging is needed, power from the engine is used to drive the generator. This eliminates the need for an external charger or power connection.

The key to the system is a power split device in the transmission that sends engine power either directly to the wheels or the electric generator. The generator, in turn, powers the electric motor and recharges the batteries. The power split device uses planetary gear to constantly vary the amount of power supplied from the engine to either the wheels or generator. The electronically controlled transmission controls engine speed, generator output, and the speed of the electric motor to handle changing driving modes. The system is designed to keep the engine running within its most efficient rpm range. When increased driving loads lug down the engine's speed, the control system shuts off fuel to the engine and kills it. The electric motor then takes over and provides 100% of the driving power. If additional power is needed, the engine is restarted and adds its power output until the extra power is no longer needed. At that point, the electric motor cuts out and the engine resumes its light load operation in its optimum speed range. The most important elements in a hybrid vehicle are shown in Figure 9.5.

Although the four companies involved in this case study had not previously worked together on the same project, they each had long-term relationships with some of the others. Company A is a U.S. car manufacturer that has purchased Company C. Company C (another Japanese firm) is an engine

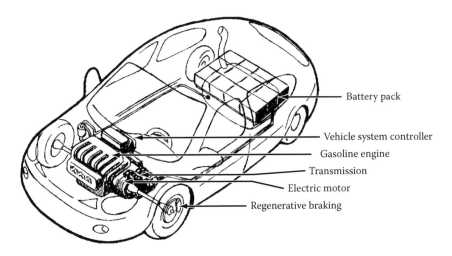

FIGURE 9.5
A sketch of a hybrid vehicle. (Courtesy of José Pérez Molías.)

manufacturer. Company B will provide a torsional damper for the project. B and C have a strong prior relationship because B long has been the main supplier of power trains for C's automobiles. HOCH-Mexico has had a direct relationship with Company C because it has provided power train components in the past. It has an indirect relationship with Company A because of its relationship to Company C. In this sense, HOCH-Mexico is far less central to the overall network than are the other firms. This relative distance will become an important factor in the development of the case study. For an easy understanding of the companies' relationships, see Figure 9.6.

This kind of network structure may seem needlessly complicated, but it is quite common in modern global organizations. It is based on the assumption that each organization in the network has distinctive expertise in developing and producing the particular item that it is contributing to the project, and

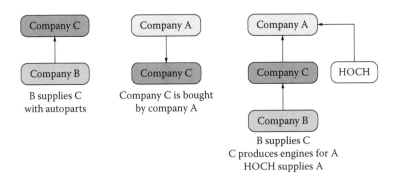

FIGURE 9.6
A, B, C, and HOCH companies' relationship.

also has the capacity to produce it at the lowest possible cost while maintaining high levels of quality and efficiency. Because it owns Company C, Company A became the central point in the network, more by default than by any preestablished plan. Because Companies B and C have direct relationships with Company A, they are in a better position to negotiate than HOCH. As a result of the negotiations, the most difficult tolerances have been assigned to the flywheel project.

Early in the project, Company B sent its design for the torsional damper to Company A, for it to be approved and forwarded to HOCH. Company B uses a different CAD platform than Company A, but has a translator that can convert its designs into a format that can be read by Company A's CAD platform, called I-DEAS™. Company A then forwarded the damper design to HOCH, evidently without verifying that it met the overall design parameters for the project, using *its* preferred CAD platform (I-DEAS). HOCH uses a different CAD platform (Pro-E™) because that is the platform used by its major customer (which is *not* Company A). However, it is not fully compatible with I-DEAS. HOCH's platform could read the design sent by Company A, but in doing so converted it from a three-dimensional model to a two-dimensional one. HOCH's engineers knew that going ahead with development based on two-dimensional designs was risky, but they were under extreme time pressure. Company A was planning to start up a new assembly line to produce its hybrid vehicle. Delays could cost the firms millions of dollars, so the company was narrowly focused on achieving results quickly, rather than on developing a successful design team and development process. HOCH went ahead with its design and produced twenty-two prototypes of the flywheel, each costing approximately US$3,000. However, when the flywheel and torsional damper prototypes arrived at the assembly station, they could not be connected.

A heated discussion ensued regarding responsibility for the design errors. Both HOCH and Company B argued that there was nothing wrong with their designs or their production processes. It was Company A's responsibility to ensure that the designs were compatible, and it had failed to fulfill that responsibility. While the debate continued, time was growing short. Initially, HOCH's engineers suggested that they simply drill some holes in the flywheel at specified places in order to join the subassemblies together and continue with the testing of the prototypes. This was only a temporary solution, because the tests found that the combined designs were not functional and were difficult to manufacture. The overall design for the automobile leaves very little space for the drive train. Each of the companies involved has an incentive to use as much space as possible because its design and production costs are higher for a small component than for a larger one. To date, HOCH has had to complete four design cycles, leading to an almost complete redesign of the flywheel. Figure 9.7 depicts several manufacturing steps for the flywheel.

Communication problems were not limited to the incompatible CAD systems. Because the design departments of the four companies are geographically distant from one another (HOCH design work was performed in Mexico), the

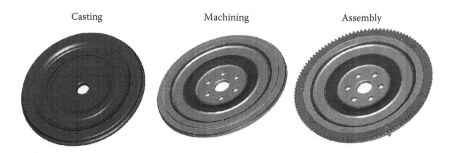

FIGURE 9.7
The flywheel from casting to assembly. (Courtesy of HOCH.)

parties who were involved relied on email to communicate. Wanting to make sure that no one was left uninformed, they copied their emails to large numbers of people in each firm, some of whom were only tangentially involved in the project. Many of the recipients replied to the emails, generating even more messages. None of the organizations assigned anyone to be a gatekeeper for all of these messages, so all of the participants were constantly inundated with information, the vast majority of which was irrelevant to their needs. As the design crisis deepened, the volume of emails skyrocketed. Buried in all of those messages was the information needed to anticipate the prototype breakdown, but it does not appear that key decision makers ever located that information. In fact, the decision-making system was so decentralized that was not even clear who was responsible for making key decisions. As a result, no one really knew who should receive a particular message. Communication among them was complicated further by almost incessant negotiations among the companies about who would be responsible for absorbing the cost overruns that were rapidly piling up. Eventually, Company A assigned a single person to manage communication about the project, but none of its employees seemed to want to play that role for very long. Within months four different people had been named to the position because the first three moved on soon after being given the assignment. Soon after taking over, the fourth decided that Company A should absorb the costs of the design errors. This eliminated the incentives that drove the incessant negotiating, but it may not have solved the underlying problems, since it meant that the other companies now might be able to shift their losses to Company A. As important, there still are no financial incentives in place that would prevent the different companies from poaching on the space allocated for the others.

Interpretation

First we analyze the GEM center ring influence (Figure 9.8).

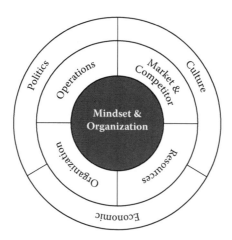

FIGURE 9.8
Global engineering model's center ring influence.

In earlier chapters we explained that the core of global manufacturing is composed of a company's global vision and the structures it creates in order to implement that vision. HOCH's global vision differs in important ways from those discussed in earlier chapters. Because it produces parts and subassemblies, rather than finished products, it always has been in the middle of complicated production chains. Consequently, it has developed a global vision that combines flexibility with centralized control. HOCH's management long has realized that becoming part of cooperative networks with other organizations can increase organizational efficiency and reduce operating costs. It is for precisely these reasons that network systems are a rapidly growing aspect of today's global economy.

Power and Politics Complicate Network Structures

Network structures carry unique complications. The most important of these involves a dilemma regarding organizational autonomy. On the one hand, cooperative networks must have decentralized control systems. If power is centralized in the hands of one party, the network takes on the inefficiencies and inflexibility that characterizes traditional bureaucratic organizations. However, if control is not centralized, each constituent organization has a large degree of freedom in how it will organize its part of the larger project, as well as strong incentives to poach resources from its network partners. As Conrad and Poole note, "This situation makes it possible for some units to take advantage of the others by 'free riding' and taking shortcuts—putting in the bare minimum that others will accept in order to maximize their [individual] profit."[4] This does not mean that it is impossible for network organizations to succeed. It simply means that members of network organizations inevitably have both incentives to cooperate with one another and incentives to compete with one another over resources, costs, and profits.[5] Those dual incentives must be managed successfully if the network is to succeed.

Communication Processes Complicate Network Structures

A second set of complications facing network structures involves communication and coordination. Traditional organizations have formal communication systems built into their structure and operating practices. They also tend to be made up of people who have long-term relationships with one another, which fosters a high level of trust. For reasons we will explain later, formal communication systems inevitably break down. However, when workers have had experience working together and have high levels of trust in one another, they usually are able to find ways to compensate for the inevitable communication breakdowns in formal communication systems. However, each member of a network organization has its own formal communication system. Unless the two organizations have cooperated frequently with one another in the past, the formal communication system of one organization is not likely to be linked tightly to the communication system of another organization. Their members also are not likely to have the kind of relational history or level of trust necessary to compensate for weaknesses in the formal systems. If the workers also have widely differing cultural backgrounds, the likelihood of communication breakdowns is increased further. Conrad and Poole conclude:

> Network organizations—with their flexible structures and complicated relationships among units—are extremely complex. This complexity ... [makes it difficult] to determine who is responsible for what in network organizations. Unless units specifically work out how they will coordinate activities and constantly communicate with each other, important things can fall in the cracks.[6]

Potential Sources of Communication Breakdowns

Processes of information exchange, both within organizations and among them, create a fundamental paradox. On the one hand, decision makers depend on receiving accurate, timely information from employees located throughout their organizational hierarchy and, in the case of network organizations, from employees in other organizations. However, if information were to flow freely through these organizations, decision makers soon would be overloaded and overwhelmed by the information they receive. For example, envision a moderate-sized hierarchical organization (one in which each supervisor has only four subordinates and the organization chart has seven levels). Each employee sends only one message a day up the chain of command. If no messages are filtered out, 4,096 messages would reach upper management each day, creating serious problems of information overload.[7] But, if every employee screens out only half of the information received, 98.4% of the information generated in the organization would never reach its decision makers. Consequently, the information flowing through formal channels, both within and among organizations, must simultaneously be distributed widely and restricted substantially.

This dilemma is complicated further by a group of barriers to effective information flow. Some of these barriers to information flow involve the formal structure of the organization and the nature of human communication. These structural barriers would exist regardless of who worked in the organization or their background, training, motivations, cultures, or competencies. When one person communicates a message to another, each of them interprets it. The words that make up the message are meaningless until some human being makes sense out of them.[8] When people communicate, they exchange their *interpretations* of information, not information in a pure form. Messages are symbols, not blueprints. When people interpret messages, they alter the message's meaning. People *condense* messages, making them shorter and simpler; people *simplify* messages into good or bad, all or none, or other extreme terms; they *assimilate* new messages so that their meaning is consistent with information received in the past; they *whitewash* messages, so that they will not upset the people to whom they are sent; and people *reductively code* messages by combining them with other information to form a sensible overall picture. In the process of interpreting a message, people simplify and clarify it. They absorb some of the uncertainty and ambiguity in the message. But, they also change it. Interpreting information is inevitable because all messages carry some degree of ambiguity, and some degree of uncertainty about how they should be interpreted. When messages are interpreted, they are changed. The more times a message changes hands, the more it is interpreted, and the more it is changed.

In addition, a number of personal and interpersonal factors also complicate information flow (see Table 9.1). Differences in power, status, motivations, and culture generate guarded communication. When people feel defensive, they tend to communicate in writing (electronic or hard copy) instead of face-to-face. Their messages focus on tasks, with little informal or social content. Written messages are more ambiguous than those exchanged face-to-face, making differences in interpretation more likely.

Communication breakdowns reduce trust, which leads employees to rely even more heavily on written communication to protect themselves, and so on in a downward spiral. When employees communicate with people in different organizations, these processes are exaggerated further. Their loyalty is to *their* organization, rather than to the network in which it is linked. Unless they have a great deal of contact with employees in the other networks, they are not likely to share high levels of trust. Supervisors can offset the negative effects by deemphasizing differences in status and power, training their

TABLE 9.1

Factors That Distort Vertical Communication

Structural	Personal and Relational
1. Processes of interpreting messages • Condensation • Accenting • Assimilation to past • Assimilation to future • Assimilation to attitudes and values • Reduction	1. Power, status differences between parties
2. Number of links in communication chain	2. Mistrust between parties
3. Trained communication • Incapacity • Perceptual sets • Language barriers	3. Subordinates' mobility aspirations
4. Large size of the organization	4. Inaccurate perceptions of information needs of others
5. Problems in timing of messages	5. Norms or actions that discourage requests for clarification
6. Problems inherent in written communication	6. Sensitivity of topics

subordinates in communication skills, rewarding their subordinates for keeping them informed, and encouraging them to seek clarification of ambiguous messages. However, they rarely are able to do these things with employees of other organizations, regardless of how tightly they are networked together.[9] Employees are especially unlikely to communicate negative information or information that deals with controversial or sensitive issues—precisely the kind of information that other employees most need to have. In highly political organizations or organizational networks—ones facing crises or involved in active negotiations over zones of responsibility or cost sharing—withholding information is even more likely, especially when it is negative. Information is a potent source of power in all negotiations, but only if it is not widely available. Political battles—among individual employees, units of organizations, or organizations in networks—often are information battles, and the side that has obtained and exploited secret information wins.[10]

Each of these factors was present to some degree in the hybrid engine case. As the problems increased and the spectrum of cost overruns grew, the situation became progressively more political. Engineering and manufacturing problems were not fully communicated up the chain of command of any of the firms, and certainly were not communicated in any formal way from one organization to the others. Structural breakdowns were complicated by problems integrating the various communication technologies, and by political, personal, and interpersonal processes. In short, the hybrid engine project quickly became a textbook case of communication/coordination problems in network organizations.

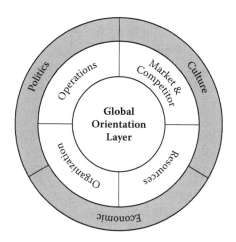

FIGURE 9.9
Global engineering model's outer ring influence.

Global Context

In earlier chapters, cultural factors were the most important element of the context within which organizations operated. In the case of HOCH, cultural differences were present, but economic constraints were more salient (Figure 9.9). HOCH's very success placed significant limits on its ability to grow. It already dominates the Mexican market, and its successful operations in the United States and Brazil prevent it from expanding into the North American and South American markets. As a result, if HOCH-Mexico is to grow or even maintain its current profit margins, it must do so by developing new products in new market niches that are related to its core business. It has already taken steps in this direction, by developing a new specialty in producing automobile parts that no longer are available through other suppliers, and by starting a line of tools that are not linked to the automobile industry. It was successful in these efforts largely because of its global vision—to develop and maintain long-standing relationships with Mexican firms, something that it had done effectively since 1975. But, those connections were relatively easy to coordinate. They involved established technologies, stable and predictable ways of doing things, and long-term, effective communication networks. This does not mean HOCH-Mexico always had smooth sailing. It had been common for HOCH to find problems in the materials provided by their suppliers. These problems included (1) machining issue due to high hardness in casting material, (2) poor quality of rolled steel, and (3) inadequate heat treatment and chemical composition of some of the materials supplied by outside organizations. Even though HOCH's global mindset calls for establishing relationships with local businesses, these quality problems eventually led it to use only German companies to supply raw materials while retaining local contacts for other materials. However, developing new technologies, as in the hybrid engine case, is a substantively different and significantly greater challenge than producing parts through established, routine processes. HOCH-Mexico was more dependent on outside organizations, their coordination problems were much more complex, and as we explained in the previous sections of this chapter, the chances for coordination problems and communication breakdowns were much greater.

The situation of HOCH-Mexico is also complicated by the economics of global competition. Many multinational corporations have located facilities in Mexico in order to benefit from its lower wage rates. Although this still is true of some parts of the country, it is not true of the area surrounding HOCH-Mexico. In part because of the success of the consortium of automobile-related organizations to which HOCH belongs, wage rates have risen significantly, so that they now are comparable to those in much of the rest of the world. During the past 4 years, some salary increases have exceeded 10% annually, and the overall wage rate is expected to increase by 7 to 8% per year in coming years.[11] Because the equipment used in HOCH's plants is complicated, the organization must hire and retain skilled workers, which is an even more expensive proposition. As a result, in 1997, the average worker at HOCH-Mexico was paid about US$9,000 per year. By 2001, the cost had risen to approximately US$16,000 per year, an increase that exceeded productivity gains.

The resulting economic squeeze has put several Mexican automotive organizations in serious trouble. Increased competition from Asian imports, which will become even more intense in the future, makes it impossible to pass costs on to consumers. If these organizations are to stay competitive and remain profitable, they must find ways to increase productivity and limit their costs, including costs that are related to their contracts with HOCH.

Because of these economic realities, HOCH-Mexico must do everything possible to maintain very high levels of production efficiency. A focus on efficiency is not new for the organization. HOCH always has been a conservative, risk-averse, and highly efficient company. It dominates its market because of that efficiency and because of its commitment to quality and excellent customer service. But, its efficiency depends on constant monitoring of internal processes, and on a strategy of outsourcing any functions that it cannot do most efficiently in-house. As a result, its profitability depends on its being able to avoid projects that lose money, and to shift excessive design and production costs to its suppliers and network partners. However, the suppliers' survival depends on their being able to find ways to avoid having costs shifted from HOCH to them, and they have every incentive to shift their costs to HOCH. In sum, for every organization involved, there is little or no slack in the system, and very strong incentives to shift costs whenever possible.

In the hybrid engine project, these economic constraints may have been exacerbated by cultural factors. Like most German firms, HOCH-Mexico is accustomed to decentralized organizational structures. These systems succeed because each work group that is involved realizes that it must keep the other units apprised of its successes and problems. Organizations that are accustomed to these systems can start to assume that no news is good news. Since the consortium also was decentralized—during the early stages of the project no one person or organization was assigned the task of ensuring effective coordination across the member companies—each company's poaching of space and costs went largely undetected. Similarly, when HOCH's Mexican engineers and workers received two-dimensional designs from Company A, they may have decided to not complain or dissent out of deference to their superiors (a key characteristic of high-PD cultures). U.S. engineers in Company A may not have inspected the design of the torsional damper closely enough because they assumed that it was only rational for Company C to stay within the constraints specified in its contracts. Mexican engineers at HOCH, who take a great deal of pride in their ability to creatively jury-rig systems in order to make them work better than they were designed to work, may have assumed that they would be able to solve any problems that would arise in the hybrid engine project. None of these factors in and of themselves seemed to be as important as the structural factors we have discussed, but they may have played a role in the communication breakdowns that led to the crisis.

Postscript on Life in Network Organizations

The situation faced by these Mexican automotive firms mirrors the instability, complexity, and environmental turbulence faced by all multinational firms competing in today's global economy. Products now change much more quickly than they once did. Product life cycles—the time from introduction to the point at which the product is outmoded or dated—have grown increasingly short. The life cycle of a computer chip has decreased from 5 years to 1. Even refrigerators have product life cycles of only 3 years now, as opposed to 10 or more a few years ago. The number of possible organizational ties, both for customers and for organizations, have skyrocketed. The range of choices for telephone or banking services, or for suppliers of drive train components, is striking.

The organizations involved in the hybrid engine network assumed that no new structures needed to be created. Perhaps because they each had worked effectively with the other partners in past projects, there seemed to be no need to change their normal ways of doing things. But, network organizations are more than the sum of their parts. Complications related to coordination and communication increase geometrically as members are added to a network, and those added complications can quickly overwhelm systems designed to deal with interactions between two organizations. Network arrangements require the creation of lateral relationships that facilitate communication and coordination across multiple organizations.

Creating Effective Integrating Structures

Over 30 years ago organizational theorist Jay Galbraith identified a number of methods organizations use to cope with the uncertainty created by rapid change and complex relationships.[12] One set of methods involve changing organizational structures. Arrayed in terms of increasing costs, these *structural integration mechanisms* include:

- Establishing *task forces*, short-term teams set up to deal with a specific problem or project. Members are drawn from several different organizations based on their special knowledge about the issue and the interests their organizations have in the project. A key challenge for the team is to overcome communication barriers posed by the fact that its members come from different organizations and have to overcome differences in experiences, terminology, and interests. If these problems can be surmounted, taskforces perform very well.

- Forming a relatively permanent *integrating team*. Sometimes a problem or project continues indefinitely or recurs regularly.

Regardless of which structure is implemented, they usually need to be supplemented by creating *managerial linking roles*. Everyone in each of the firms that are part of the network must agree that the project is important, and upper management in each firm must agree to back up the recommendations and actions of the new structure and the managerial links. To give the integrating team more legitimacy and power, its formal manager should be directly connected into the hierarchy of authority and have control of the project's budget. He or she represents the team to each organization and serves as a symbol. Having an integrating manager gives the team more legitimacy and resources to work on its own. Although being an integrating manager can be stressful, playing the role can enhance a person's career because it broadens the liaison's outlook and sharpens his or her communication skills. Reporting back to their home organization/department helps members keep their skills sharp and keep up to date on the latest developments in their fields. Other members of this team keep the engineer from applying only the perspective of his or her home organization/department to the problem, and he or she will keep other team members honest by making sure that issues important to his or her organization/department are considered each time the team makes a decision. Upon return to the materials department, the engineer is able to consult with others about the relevant problems encountered during the project, which makes him or her an even more valuable member of future project teams.

Organizations usually choose among these integration mechanisms based on a simple cost-benefit analysis. This evidently was the primary motivation behind the hybrid engine consortium's decision to rely on existing communication networks. Benefits of the structural mechanisms depend on each method's effectiveness in handling the level of uncertainty the organization experiences. Because the hybrid project was so new, it involved levels of uncertainty that exceeded the capacity of these existing structures. Liaisons and task forces are effective for moderate to high levels of uncertainty. Organizations facing high levels of uncertainty will be more effective if they utilize more expensive structures, including integrating teams and managerial linking roles. But costs must also be considered: these include personnel expenses, time spent learning to use the method and getting it to work smoothly, information load imposed by the method, and the amount of stress members experience. The hybrid engine consortium's decision to create a managerial linking role and the turnover it has experienced among employees assigned to that role illustrate both its importance and the personal costs of the system. Eventually the consortium is likely to shift to an even more expensive system, for example, an integrating team, in order to distribute the stresses more widely.

One company that designed integrated computer hardware and software systems had several teams working on different parts of its new System Alpha. To deliver System Alpha on time and in shape, it was vital that each team's work be compatible with that of the others. The company created an Alpha management team whose assignment was to coordinate the work of the System Alpha teams. An engineer who had informally worked on coordinating the teams involved in System Alpha was appointed manager of the management team. She was given a budget for personnel to test the compatibility of products that the different teams were creating and to provide extra help for teams having problems. This extra heft that the manager had gave her the ability to stimulate the teams to action, and System Alpha was delivered on schedule.[13]

Creating Effective Communication Technologies

Electronic technologies like email, computer, audio- and videoconferencing, blogging, and instant messaging have become absolutely essential to the operation of modern global organizations. However, as the overuse of email in the HOCH case has illustrated, unrestricted use of those technologies can create more problems than they solve. These technologies typically increase linkages both within and across units and organizations at all levels. They aid in information flow and in the formation of informal relationships. But they are most valuable when combined with one of the structural mechanisms that we just described, as when a liaison uses email or a computer conference to maintain contacts with the units or organizations he or she is responsible for.

And, they are most valuable if used in conjunction with an appropriately designed *knowledge management system*. Whereas work integration systems link people and units through the common work they perform, and communication systems integrate people by enabling contact and communciation, knowledge management systems link people by building a common knowledge base that they can draw on in their work.

Intec, an engineering and project management firm that serves the petroleum industry, is one example of how integration through knowledge management can be done.[14] In 2002 the company found that it was becoming more and more difficult to keep track of and access information and knowledge that had been accumulated over years and many projects. The company formed a learning team to put together a knowledge management system that would compile and make useful knowledge available so that Intec engineers could learn from one another's experiences and weren't always reinventing the wheel when they undertook a new project. The learning team put together a database system that integrated existing knowledge resources such as manuals and previous bids, automatically located experts on various topics, facilitated the identification of best practices, and captured information from engineers' work automatically, all with an easy-to-use interface. While this sentence makes it sound easy, it took the learning team over a year to do this, and they had the assistance of an excellent knowledge management vendor. The resulting system enabled engineers to capitalize on the firm's knowledge and saved the company over $200,000, not to mention improving the quality of its work. Obviously, knowledge management systems will work most efficiently if all interested parties have access to the system and possess the technologies (and software) necessary to use it effectively. Initially, purchasing a three-dimensional CAD system seemed like an excessive expense for HOCH-Mexico; in retrospect, it seems like a pretty wise investment. To solve the data exchange problem, some companies are using the same CAD system as their client, the objective being to talk the same language. However, not all the companies are in the position to invest in a new CAD system, which demands money, effort, and time to give productive results. This problem is aggravated if one considers that one Original Equipment Manufacturer (OEM) is not only serving one assembly plant, but several clients with different CAD systems, and no company is in a position to install more than one CAD system. To solve this problem, there are specialized consulting services that offer data exchange utilizing neutral interfaces such as the Initial Graphics Exchange Specification (IGES), Standard for the Exchange of Product Model Data (STEP), Drawing Exchange Format (DXF), and others. These consulting companies are also waiting for the development of powerful interfaces that will allow the transferring of not only geometry, but also other characteristics of a model, such as parameterization, which is becoming a great need for engineering design changes, in an efficient manner.

Power, Politics, and Incentives in Network Organizations

How do network organizations motivate the members of individual units to collaborate and be innovative? This problem cannot be solved by a leader's dictates, because the flatter network organization does not have hierarchical authority over its constituent units or their members. Indeed, hierarchical authority would undermine flexibility and adaptability, which are the greatest advantages of the network form. Another challenge arises from the fact that the network gives units a degree of freedom in how they organize their work and the effort they put into the larger enterprise. This situation makes it possible for some units to take advantage of the others by free riding and taking shortcuts so they maximize their profit by putting in the bare minimum that others will accept.

Network organizations can motivate and control their units and their units' members through three complementary routes. First, they can attempt to cultivate trust in the network. Trust is the ideal cement for the network organizations. They have little or no hierarchy, so hierarchy cannot be the source of authority to coordinate and control. Moreover, networks are often composed of numerous different organizations, each with its own culture, so culture cannot be the basis for control. Trust is a special property of the relationships among members of the network that enables them to act on the assumption that others will fulfill their own responsibilities in good faith. It is achieved through engaging in cooperative action with others and through observing their competence and their willingness to live up to their commitments. To cultivate trust, the managers of the new product network might ask the programmer to work in the field for a couple of weeks with programmers from other components of the network. If their work goes well, programmers from the various firms come to trust each other and carry back good reports to the rest of the network. Trust will also be built in this network as various units carry out their responsibilities on time and effectively.[15] For the hybrid engine network, building trust will be a very difficult undertaking because every party is well aware of the potential for free riding. But, it is crucial for the success of the project.

A second source of motivation in network organizations is an inspiring, meaningful task. Network organizations are typically task or product focused, and this gives all members a common frame of reference. A meaningful task or goal can inspire the units and individuals in the network to work hard and ensure that they coordinate with other units. For example, the goal of the networked floral company Calyx and Corolla is to deliver the best possible flower arrangements faster than any other company and at a good price. This goal unifies the efforts of each part of the network: the flower growers try to deliver high-quality flowers; the arrangers work out innovative designs; the express delivery companies configure their operations to make sure the flowers get there fresh; and the coordinating unit, Calyx and Corolla itself, works hard to make sure that the parts of the network connect smoothly with each other as the process moves from flowers in the field to flowers on the table. As far as we can tell, the hybrid engine consortium did little to create the feeling that their project was important, either to the economic realities faced by the companies that were involved or as an exciting project in itself.

A third source of motivation and control in network organizations is network-based formal systems for monitoring and control of members and their activities—the purpose of the managerial role created by Company A during the latter stages of the hybrid engine project. In addition to trust, networked organizations may also attempt to develop structures to formally coordinate unit activities. These systems are based on contracts among the units in the network that provide a formal, written understanding among the units concerning their responsibilities and compensation.

A final source of control and coordination is for management in all network organizations to focus on not playing blame games. In complex systems, like network organizations, it is often impossible to determine the cause of problems. Consider a relatively simple organization composed of three units: one that designs widgets, a second that produces them, and a third that markets and distributes them. The marketing unit may find that there is not enough of product X to meet the demands of an important customer. Its initial tendency is to blame production, which has immediate responsibility for making X. But production may be having trouble due to a design problem that causes a part of the product to break when it is removed from the stamping presses. The blame then seems to shift to design. However, the design unit used the flawed plan for product X because it had not gotten any feedback from production on the problem. Moreover, design understood from marketing that customers really appreciated the part of X that tended to break off in production's machines, so they wanted to keep it. Does the problem then trace back to production's lack of feedback or to marketing's insistence that the part of the design that caused problems be retained? The answer is that none of these can be said to be the sole cause of the problem. Causality in networks is ambiguous. The problems persist because of the organizational system as a whole, a system in which design does not communicate with the other units, in which production is not particularly proactive about problems it encounters, and marketing is out there selling stuff without considering whether other departments can meet the delivery schedules it sets. In cases where it is difficult to determine the causes of problems, it is also difficult to solve them. When eliminating the source of a problem means changing the entire system, the problem may recur, because systems change slowly at best.

Review and Study Questions

1. Explain how a hybrid vehicle operates and how it manages energy.
2. Investigate which automotive companies are selling hybrid vehicles nowadays. Give a short explanation of their vehicles' characteristics.
3. What are the priorities of German engineers in their professional life?

4. Fill out the following table to clearly show a summary of differences and similarities of German and Mexican working cultures.

German-Mexican Culture Summary

Differences		Similarities
Germany	Mexico	Germany and Mexico

5. Provide measures to take advantage of culture similarities while working with Germans and Mexicans in global engineering projects.
6. Provide measures to overcome culture differences while working with Germans and Mexicans in global engineering projects.
7. Explain the problems arising when exchanging technical information among different CAD platforms.
8. Explain the relationships among the different companies involved in this case study.
9. Explain the coordination and communication of cooperative international networks. Explain the main characteristics of network organizations.
10. Explain how good communication can be achieved among distant working networked subjects.
11. Investigate the average medium (incoming, 5 years experience, 10 years experience) salary for engineers in Mexico and Germany. Conclude on the differences found.
12. What is a knowledge management system? Investigate some knowledge databases of your local companies.

Notes

1. Charles Fauchex, "Strategy Formation as a Cultural Process," *International Studies of Management and Organization* 7 (1977): 127–138; R. A. Friday, "Contrast in Discussion Behaviors of German and American Managers," *Journal of Intercultural Relations* 13 (1989): 429–446; H. K. Fridrich, *A Comparative*

Study of U.S. and German Middle Manager Attitudes (Boston: Sloan School of Management, MIT, 1965); J. H. Horovitz, *Top Management Control in Europe* (London: MacMillan, 1980); A. Kieser and H. Kubicek, *Organisation* (Berlin: Walter de Gruyter, 1983).

2. G. W. Preiss, "Work Goals of Engineers: A Comparative Study between German and U.S. Industry," unpublished master's thesis (Boston: Sloan School of Management, MIT, 1971).

3. See Geert Hofstede, *Culture's Consequences*, 2nd ed. (Thousand Oaks, CA: Sage, 2001), chapter 3.

4. Charles Conrad and M. S. Poole, *Strategic Organizational Communication*, 4th ed. (Ft. Worth, TX: Holt-Rinehart, 1995), p. 212. Also see Brian Uzzi, "Organizational Networks, Structural Embeddedness, and Firm Survival," paper presented at the national meeting of the Academy of Management (Dallas, TX: 1994); and Keith Proven and H. Brinton Milward, "A Preliminary Theory of Interorganizational Network Effectiveness," *Administrative Science Quarterly* 40 (1995): 1–33.

5. For an excellent example of a successful network organization, see Larry Browning, Jan Beyer, and J. Shetler, "Building Cooperation in a Competitive Industry," *Academy of Management Journal* 38 (1995): 113–151.

6. Charles Conrad and M. S. Poole, *Strategic Organizational Communication*, 6th ed. (Belmont, CA: Thomson Wadsworth, 2005), p. 216.

7. This is why computerized management information systems, recently installed in virtually every major organization, have had perplexing effects. Computer information systems do not filter information. In theory, they allow every employee, no matter where in the organization, to instantly access any part of its information base. However, no one can process all the information. Unfiltered formal communication will literally bury upper-level managers in information, at least until they learn to use the equipment to screen out messages. High-speed computer systems may only allow them to be buried more quickly. The solution to the problem of communication overload is for upper management not to use the systems, which defeats the purpose of installing them in the first place. See Ron Rice and Urs Gattiker, "Communication Technologies and Structures," in *The New Handbook of Organizational Communication*, ed. F. Jablin and L. Putnam (Thousand Oaks, CA: Sage, 2000). Also see Fredric Jablin, "Formal Organizational Structure," in *Handbook of Organizational Communication*, ed. Fred Jablin et al. (Newbury Park, CA: Sage, 1987).

8. Eric Eisenberg, "Ambiguity as Strategy in Organizational Communication," *Communication Monographs* 51 (1984): 227–242.

9. Cal Downs and Charles Conrad, "A Critical Incident Study of Effective Subordinancy," *Journal of Business Communication* 19 (1982): 27–38; Gail Fairhurst, "Dialectical Tensions in Leadership Research," in *The New Handbook of Organizational Communication*, ed. F. Jablin and L. Putnam (Thousand Oaks, CA: Sage, 2000).

10. Janet Fulk and Sirish Mani, "Distortion of Communication in Hierarchical Relationships," *Communication Yearbook 9*, ed. Margaret McLaughlin (Newbury Park, CA: Sage, 1986).

11. Study done by Intergamma and Banxico (Bank of Mexico) and published in *Reforma*.

12. Jay Galbraith, "Organizational Design," in *Handbook of Organizational Behavior*, ed. J. Lorsch (Englewood Cliffs, NJ: Prentice-Hall, 1987).

13. Richard Daft, *Organization Theory and Design*, 3rd ed. (St. Paul, MN: West, 1989).

14. Kathleen Melymuka, "Smarter by the Hour" *Computerworld*, June 23, 2003, pp. 43–44.

15. Dale E. Zand, "Trust and Managerial Problem-Solving," *Administrative Science Quarterly* 17 (1972): 229–39. Also see Peter Monge and Noshir Contractor, "Emergent Communication Networks," in *The New Handbook of Organizational Communication*, ed. Fredric Jablin and Linda Putnam (Thousand Oaks, CA: Sage, 2000).

Section IV

Case Studies: Applying Concepts

10

Technical Consulting and Organizational Recovery at CHips

CHips is one of the world's largest suppliers of semiconductors and related services. It is headquartered in the southwestern United States, where much of its high-end design work is done. Founded in the late 1960s, it pioneered outsourcing of integrated circuit (IC) assembly and testing services to plants throughout Asia. Although outsourcing has been controversial in the United States, and is becoming controversial in other parts of the world, from its beginning CHips' management viewed it as the only viable way of implementing its goal of providing ICs that are smaller, lighter, faster, and cheaper.

CHips focuses on all three phases of IC: fabrication, assembly, and testing. Fabrication involves depositing transistors and circuitry on silicon wafers—assembly (which also is called packaging in the industry) is the process of cutting silicon wafers into individual ICs and placing them in protective housings that provide the electrical interconnections between the ICs and system boards—and testing to ensure the proper functioning of CHips operating in rigorous daily use. In addition to research facilities in Europe and the United States, CHips currently has factories in six countries—South Korea, Japan, China, Taiwan, Singapore, and the Philippines—and employs more than twenty-four thousand people working in more than 5 million ft^2 of floor space in the key microelectronic manufacturing centers of Asia. It has sales and service centers throughout the world and is a partner for more than two hundred of the world's leading semiconductor companies. It anticipates revenues of more than US$50 billion in 2008.

However, CHips' growth has not been steady or stress-free. Like the entire semiconductor industry, the company experienced rapid growth during the 1990s, suffered through the collapse of the dot-com bubble in 1999–2000, then steadily recovered so that by 2005 its revenues exceeded those of 1999. Part of the reason for the recovery is its management's commitment to being proactive—anticipating industry trends, constantly looking for partnerships that will be stable and productive over the long run, and abandoning product lines or production facilities that no longer contribute to its growth. The biggest challenge in being proactive is being careful and strategic—not acquiring facilities or product lines just in order to grow, and not abandoning facilities or lines prematurely, without carefully assessing their strengths and weaknesses in both the short and long term. We know now that CHips' management made the right decisions.

But, by the late 1990s, CHips' management was worried. The semiconductor industry was rapidly becoming more and more competitive, and more and more global. While some of the company's facilities were state of the art, in terms of both the technology being used and the ways in which they were being managed, there was a great deal of inconsistency across those facilities. There was simply too much slack in the system. To make matters worse, the company's internal system of communication and control had not kept up with its growth. Much of the time headquarters lacked accurate and up-to-date information about exactly what was going on in some parts of its far-flung operations. Management had been so busy seeking out and exploiting new opportunities that it had not had (or had not taken) the time to step back and evaluate the overall system. Fortunately, they realized both that there were weaknesses in the system, and that they were too close to the overall strategy and too isolated from everyday operations to objectively analyze their strengths and weaknesses. In short, the company needed a "new set of eyes" to determine where the inefficiencies were located, what caused them, and what could and should be done about them.[1]

One member of the management team suggested that they contact two of his former engineering professors and see if they would be willing to provide the needed expertise. After some informal negotiations, the parties developed a multistage consulting contract. This case study tells the story of that collaboration. At one level it will focus on the development of technical solutions to complex technical problems; CHips' management charged the consulting group with developing a technical solution to its logistical problems that could be used in all of its facilities, regardless of where they were located or what they produced. But it also tells another story, about the unique complications that occur in multinational operations.

The Paradox of Cross-National Operations

Truly global corporations face a number of unique challenges. The most important of these is a paradox of flexibility and consistency. On the one hand, each of the company's facilities exists in a distinctive cultural, political, and economic context. This diversity requires flexible and situationally appropriate systems and practices. Geert Hofstede observed that "the structure and functioning of organizations are not determined by a universal rationality. Making shoes, producing electricity, or treating the sick in Germany as opposed to France calls for organizational structures and processes that differ in several respects."[2-4]

On the other hand, the various operations of a global company must be coordinated in a way that maintains a high level of efficiency and effectiveness in the organizational system as a whole. This need for coordination necessitates at least a minimal degree of centralized control from the home office, and at least an adequate degree of consistency in the operations of every part of the organization. Globalization expert Barbara Parker summarizes this paradox clearly and succinctly:

> Global organizations take shape by borrowing best practices and new ideas from among multiple cultures, but they must at the same time create an internal culture that instills unity and provides direction. Common worldwide processes and structures often are introduced for this purpose, but human decisions to aid or subvert a new process or structure make or break the global organization as it struggles to learn and grow a culture responsive to the dynamics of global change.[5]

Of course, there are a number of different strategies that corporations can use to deal with this paradox. The traditional approach, and the one that seems to be preferred by U.S. firms, is a global or transnational approach.[6] Foreign operations are linked to the central office like the spokes of a wheel are related to its hub. Each operation is linked directly to the home office, with little or no direct contact with the other operations. Organizational decision making and learning are centralized in the home office, upper management typically includes few foreign employees, and the organization's focus is on the consistency pole of the flexibility-consistency paradox.

An alternative multinational or international approach breaks with the notion of an organization as spokes of a wheel. Foreign operations have active and legitimate relationships with one another, based on their customers' needs and their supply chains and marketing processes. Top management teams are a microcosm of the entire corporation, and headquarters functions more as a consultant or coordinator than a policeman or dictator. The company consists of a set of centers with special expertise on different topics. When the organization faces a particular challenge, the center with the greatest expertise takes charge; when the challenge changes, so does the organizational structure. The hierarchy of the traditional bureaucracy is replaced by a heterarchy based on recurring interdependencies.[7] For example, customers in South Asia repeatedly make demands on CHips' Singapore operations. Leaders there find that some of the technical expertise they need in order to respond to these demands is located in the company's Taiwan units, and the best place to adapt the company's designs to the South Asian market is the Philippines. By repeatedly cooperating with one another, the three units came to understand one another, develop effective communication networks, and eventually began to depend on one another. The role of corporate headquarters is to coordinate these arrangements, in order to make sure that no one unit is overburdened by its many contacts with other units, and no unit is left underutilized. Fons Trompenaars and Charles Hampden-Turner list Shell Oil, IKEA, Ericsson, and Proctor & Gamble as companies that have successfully used this approach.[8]

When times are good and an organization has plenty of resources at its disposal, the flexibility-consistency paradox is relatively easy to manage. In tetrarchies, each unit can develop its own communication links with the other units, learn their strengths and weaknesses, and predict which ones would be the best collaborators on new projects. This is the essence of the strength of weak ties concept discussed earlier in the book.[9] In hub-and-spoke systems, there is time for all of this information to be collected by headquarters and appropriate decisions made. But, when time and resources are scarce and competition fierce, coordination can become a critical challenge.

Does Technology Transcend Culture?

CHips' management assumed that their coordination and efficiency problems could be solved through the use of improved technology. No one seriously doubts the importance that technological change has had on processes of globalization or the ways in which organizations operate in a global economy. The debate, to the extent that one still exists, involves the relative importance of technological and cultural factors and processes. But, this way of bimodal thinking—either technology or culture is most important—is itself a cultural construction. Western industrial societies are more supportive of either/or thinking; others teach their members to think more in both/and terms. Trompenaars and Hampden-Turner take the latter perspective: "Technologies have a logic of their own which operates regardless of where the plant is located. Cultures do not compete with or repeal these laws. They simply supply the social context in which the technology operates."[10] If the same technology is introduced to operations in many different cultures, it actually may begin to reduce cultural differences. Technologies are a form of knowledge, and when employees in different operations share the same knowledge base, they begin to think more alike over time; they experience knowledge isomorphism, to use the academic terms.[11] But, it takes time for this convergence to take place. In the early stages after a new technology is installed, its use will depend on the cultural, political, and organizational cultures in which it is introduced.

The Facts of the Case

It was early 1999 when the team received a phone call from one of their PhD students who was co-oping at CHips' world headquarters in the southwestern United States. The so-called dot-com boom was at its peak. Like the rest of the industry, CHips had been growing rapidly, acquiring new product lines and new production facilities across Asia. But, the growth was so rapid that the home office had started to feel that they were losing control of day-to-day operations. Every plant was asking for new equipment, both to keep up with technological advances and in order to cope with skyrocketing demand. But, the company lacked a systematic way of assessing how efficiently different plants were using the resources they already had. When headquarters asked for information, it was slow in coming, and given the little bit of information that was available, upper management had a feeling that there were wide discrepancies across different plants, in terms of both how efficiently equipment was being used and how rapidly different plants learned to use new equipment. Since engineers aren't all that comfortable with running multi-million-dollar operations based on feelings, the senior vice president decided to accept the PhD students' advice to explore having his professors give CHips a new set of eyes to make sense out of their situation.

As often is the case with multinational operations, finding a time and place to get the team and the VP together was a challenge in itself. Eventually they decided to meet at an airport and talk between connecting flights. At first it was chaos—flights were late, connections missed, and so on—but eventually they were able to sit down and talk. As the VP spread a massive blueprint of the company's far-flung operations out for the team, he made it clear that he respected their expertise: "You guys are professionals; you're on the cutting edge; you teach this stuff every day." And, he knew he was going to have to do *something*, and that whatever steps he took would be accepted more readily by his plant managers if they bore the stamp of an American academic team. Time went by far too quickly, and the meeting broke up with a lot of questions left unanswered. But, through emails and phone calls, a clearer picture started to emerge. Although upper management eventually wanted to create a centralized capacity planning system, in the short term they just wanted to know what was actually going on in two of their newest acquisitions, a group of plants in the Philippines and in South Korea. Their hunch was that the Philippine operation was much less efficient than most of the company, in part because its buildings and technology were older, and in part because they had acquired it as a joint venture along with another large, but recently declining U.S. high-tech firm. So, CHips offered the team an initial consulting contract. Their charge was to get back on an airplane and go visit the new operations. The VP tried to not bias the team, but the message he left them with was to "be tough; whip them into shape."

Phase I

In some ways, the VP's hunch about the Philippine operation was correct. The buildings were old and had been converted from other kinds of operations. Moreover, some of the operations were archaic. For example, instead of having a modern decision support system, the plant relied on a bunch of engineers sitting around a table making cost and production computations on hand calculators. In a way, they were just as blind as the U.S. headquarters—no one had timely access to the information they needed to make efficient decisions. And, the communication breakdowns were a two-way street—the plant managers had to make decisions without timely cost information from headquarters. However, once the team sat down and talked with the Philippine engineers, it was clear that they were pretty skilled. They were aware of modern techniques, but did not have access to them or know how to implement them if they gained access. Other inefficiencies resulted from Philippine culture. The production workers took breaks and lunch together. It was a time for camaraderie, for talking, walking around, and smoking. While on break, the plant came to a grinding halt, the machines left unattended and inoperative. There was one other surprise. The first language of the team was Spanish, while the first language for the Philippine managers and workers was Tagalog. Although formal conversations in English went well, the locals often had side conversations during regular formal meetings that the team members simply could not understand. The team noticed that the side conversations in Tagalog during meetings were not considered to be inappropriate or unprofessional. Moreover, after a few minutes of a side conversation, managers did not translate the side discussions for the team. Situations like this made the team realize that there was a lot going on under the surface. However, their charge had been to get a general sense of what was going on in the Philippine operation, and they had learned enough to meet that goal. They got back on the plane, this time to South Korea.

One of the Korean plants was a large multistory building in the center of town; the other was on the outskirts. The team was immediately struck by how big and how empty the buildings were. They also were struck by the deference they were shown. The VP was almost a god to the Korean managers and workers; as his representatives, the team members were almost demigods. Although the Korean plants were much more modern and more tightly organized, there were some similarities. Like the Philippine engineers, their Korean engineers were making complex calculations by hand, but they were also using desktop computers and Microsoft Excel instead of hand calculators. Lunch also was a time for socializing, but with a distinctively Korean twist—after quickly consuming their food, the Korean staff went to a downstairs room for karaoke. Since both of the team members were professional musicians, this ritual did not present any problems for them, but it was a new and different experience, to say the least. Even with karaoke, the plant manager seemed to be distant, maybe even suspicious of the team.

The tension may have been because of simple language problems. It simply is not true that the whole world speaks English, especially if they are from small towns or are lower-level employees. Some misunderstandings were cultural, but in unpredictable ways. At one point, a plant manager invited the team to attend a Protestant Christian revival service, an invitation that the team politely declined. They never really figured out the reason for this choice of entertainment for foreign guests—it could have been because that is how Korean managers spend their leisure time, or it could have been because they (incorrectly) assumed that all Americans enjoy revivals, or it could have been for any number of other reasons.

Other events were predictable. A key component of Asian culture is a complex mixture of business and interpersonal relationships, with the latter being necessary for the former. The tension between the team and the Korean managers continued until one cold, snowy, slippery night when the group went out to buy toys for the team to take home to their children. The shopping trip got sidetracked to a local bar, where everyone enjoyed drinks—probably too many—with one another. One of the most interesting moments of the night happened when one of the consultants slipped in the snow and lay on his back while the Korean manager looked at him and happily shouted, "This is what you get for drinking too much." At the end of the evening (or rather, in the wee hours of the morning), the Korean manager arranged to have a cab take the team home. No business was conducted (and no toys were purchased), but from that night on, the working relationship was much better.

Phase II

It was obvious to the team that their report had to focus on concrete steps to improve efficiency. Fortunately, there were obvious ways to improve efficiency. For example, in every plant, work was arranged almost randomly rather than through any kind of systematic production flow analysis layout. Machinery was located because of its access to electrical plug-ins, stairs, doors, or elevators, not because of where it fit in the production process. Redesigning workflow and the placement of equipment could speed things up substantially. Second, CHips needed to develop an electronic database management system, both to replace the inefficient manual systems in the two countries, and to better connect the plants to headquarters. Since the company's operations were unique, new software would have to be written for this system, and since the Philippine workers lacked the high-level technical know-how to operate the system, headquarters would have to provide some remedial training to prepare them for the new system. Third, production systems needed to be redesigned in order to reduce downtime. At times 70% of workers were idle while awaiting materials, repair or setup of equipment, or other workers to return from breaks. This was more of a problem in the Philippine plants, and was clearest during lunchtime. Workers needed to be trained on how to set up their own equipment, perform routine maintenance and repairs, and be cross-trained so that they could fill in when another operator was missing. Space allocation needed to be systematically examined, and lines of authority and responsibility needed to be clarified. The team concluded that before the plant managers' requests for more employees or equipment were granted, they needed to significantly increase the efficiency of their current resources through the development and implementation of a formal capacity planning system.

After examining the team's report, CHips' management decided to accept their recommendations, and offered them a long-term contract to design and implement the necessary changes. The first step was to hire a software engineer to develop the new technology. The best person for the job was a friend in Peru, who both team members trusted and who had years of experience as a database programmer. The second step was to return to the Philippines and conduct training sessions designed to bring the professional staff up to speed. While there, the team installed a system of staggered breaks/lunches that was designed to reduce unnecessary downtime. In many ways, this visit was frustrating. Although the training sessions were designed to start at a basic level and work upwards, the team got the distinct impression that the material was over the heads of the Philippine engineers.

The plant manager was clearly unhappy. He felt that the team's report placed too much of a burden for change on his operation, and was not pleased that headquarters used it to put his requests for new workers and equipment on hold. After greeting the team on their return, he basically disappeared. The new break system never really worked in the Philippine operations—using breaks for socializing was simply too deeply embedded in Philippine culture to be overcome by appeals to efficiency. Fortunately, the team was able to hire a recent engineering graduate from Arizona State University to be an on-site contact in the Philippines and continue the training and redesign process.

Things would go better in Korea, or so the team thought. After all, the staggered break/lunch system had gone beautifully. Workers lined up, moved quickly through the cafeteria, and then went back to work. Fewer people in the basement at any one time meant that more people got to sing. The software design was going well, and one of the students who accompanied the team on their second visit to Korea decided to stay on and serve as a liaison between the team and the Korean plants. He was carefully selected—a native Korean, he knew the culture and the language and he was an expert in the technology. The team could go home knowing that he would keep them apprised of what was going on and what still needed to be done. Unfortunately, he never was accepted by the Korean engineers and managers simply because he was a student—low in the prestige hierarchy in a culture that values position very highly.

Implementing the new design was a similar story. Even though the team was linked to Asia through email, telephone, Internet conferences, and occasional face-to-face meetings, there were many misunderstandings and communication breakdowns. Although today's Internet conferencing technology might have prevented many of these problems, the team still would have to face the simple realities of time and space. Implementation involved a very high level of technical detail, especially of the decision support system. The team was separated from headquarters by 1 hour and 1,000 miles; they were separated from the plants by half a day and half a globe. Crisis management meant sudden conference calls at 2:00 a.m. the team's time, and unscheduled team meetings were called at 4:00 a.m. in the middle of the Asian workday, when the plant engineers encountered an unexpected problem. This kind of chaotic scheduling is inevitable for companies with global operations.

For example, India's biggest Internet technology company (TATA) recently moved much of its engineering operation to Uruguay. U.S. multinational corporations had been overloading TATA's third shift, the one that most of its experienced engineers tried to avoid because it forced them to be separated from their families. By moving many of its Indian engineers to South America, with only an hour or two time gap, they could coordinate operations with their U.S. customers while still retaining a labor cost advantage.[12] Like TATA's engineers, the team was getting exhausted at the same moment that the Asian engineers were getting most panicky.

Eventually, the team shifted more and more responsibility to their contacts in Asia, as much out of a need to maintain their own sanity as to guide implementation. This gave them a sense of relief, but more surprises lay ahead. In spite of all the resistance, implementation actually went pretty well in the Philippines. Their engineers took the software, learned how to use it, and implemented it almost exactly as the team had envisioned. In contrast, the Korean engineers excitedly accepted the new software, thanked the team for developing it, and immediately started adapting it to better fit their operations. At one level this should not have been a surprise—Koreans are known for their ability to creatively adapt Western technologies to fit their needs, and research on communication technologies consistently find that they are implemented in novel and surprising ways—but it still was frustrating to see them alter the product of all those sleepless nights. Plus, the purpose of the entire project had been to create consistency across all of CHips' operations, and every adaptation frustrated that goal.

The second surprise came from a completely external source: just after the turn of the century the worldwide IT/dot-com burst. CHips had sought the team's help because their business was booming and they needed to improve efficiency in order to cope with the demands of growth. Suddenly, managing growth was no longer the issue, but the new efficiency was necessary to survive in a plummeting market. In 2000 the semiconductor industry was almost 50% larger than it had been in 1995; in 2001, it was smaller. CHips' fortunes followed the industry curve. The team's key contacts in Asia disappeared, the victims of corporate downsizing. But, the systems they had developed, and the efficiencies they had created, remained. When the industry started to recover in 2003, and the company diversified further, they were prepared and prospered. Still, the team had learned two important lessons: time and space are real, and designing systems at a distance is much easier than implementing them.

Interpretation

A career as a consultant often sounds attractive to college students and young professionals. Being able to fly around the world using one's hard-earned cutting-edge technical expertise to seemingly unsolvable problems is about as close to being James Bond as a young scientist/engineer is likely to come. Getting paid well only adds to its attractiveness. But, consulting is far more complicated than it seems on the surface, and the emergency 2 a.m. meetings experienced by our team made it clear that there are a number of downsides to the consultant's life. Understanding those complications starts with an understanding of the different forms of consulting.

Forms of Consulting

The most common, and perhaps the most lucrative, type of consulting involves the application of predesigned packages to what may be unique organizational situations. In its most extreme form, package-based consulting creates a patriarchy that may do no more than perpetuate managerial or technological fads.[13] The consultants' specialty—whether it is a software package, auditing systems, production flow diagrams, Total Quality Management (TQM), or organizational development (OD)—is the focus of attention, not the organization and the specific challenges that it faces.[14] Professionals' credibility, and their authority over client companies, stems from their outside status and their image as unquestionable experts.[15–17] That image depends on the consultant's expertise, but it often is even more closely related to his or her ability to maintain an image of superiority, which may have more to do with his or her ability to "translate the client's desires into a professional metalanguage and/or explain to the clients what is possible" and what is not possible in their situation.[18] Clients are persuaded that their problems result from not having bought what the consultant is selling, or that a complex web of challenges can be addressed through a single technique. The strength of this approach is that it draws on the consultants' expertise. But there are a number of potential weaknesses.

The most important potential weakness of package-based consulting is that it ignores underlying organizational politics. Why do organizations (or, to be more accurate, upper management) hire consultants? The obvious answer is that they have expertise that no one in the organization has. In *Power in Organizations*, management scholar Jeffrey Pfeffer noted that "most organizations have within them the knowledge and expertise necessary to solve their own problems."[19] In fact, if a consultant was to propose a solution that was so different than what the organization already knew, it would not be taken very seriously. Much of an organization's knowledge may be buried in routines or individuals' experiences, so if consultants ask the right questions, they can uncover expertise that the organization is not using.[20,21] But, package-based consulting is based on dictating solutions, not asking questions. In fact, uncovering existing expertise would undermine the credibility of the outside expert.

What outside experts can do is affect the internal politics of a corporation. By defining problems in certain ways, and proposing certain courses of action to solve them, they increase the organizational power of individuals or coalitions who perceive reality in certain ways, and reduce the power and influence of individuals/coalitions who see their organizational worlds differently. In a sense, outside consultants break ties created by the internal politics of an organization. Over time, certain ways of viewing the world are legitimized and become dominant; other views are discouraged or eliminated. Over time, the only consultants who are hired are those who define situations in the same way that the dominant coalition of an organization defines them, and the only solutions that are seriously considered are those that are consistent with the consultants' expertise: "the outside expert becomes part and parcel of the contest for control and power that occurs within organizations." The fresh eyes that management seeks are not really fresh at all. Finally, when the organization has multinational operations, package-based consulting has an additional disadvantage—it can sacrifice flexibility and cultural adaptability for consistency.

Fortunately, there are forms of consulting that focus on the client organization and on adapting the consultants' expertise to its unique needs instead of on imposing an existing consulting package. These relational approaches rely on developing a personal relationship with particular organizational clients. Their goal is to empower the client, not the consultant. They focus on asking questions and listening, and on learning from employees throughout the client organization, not on telling the client how to use information provided by the consultant/expert. Relational consulting takes more time and energy than package-based consulting, and its success depends on the consultant's ability to gather information from throughout the organization and create networks through which that information is shared. This does not mean that the consultants' technical expertise is unimportant. They are able to draw upon past experience, contacts, and expertise to devise new solutions for new organizational challenges, for example, provide an excellent example of engineers drawing on their past experience with using ceiling fans to cool production facilities to cool personal computers.[22] Their technical expertise was relevant, but as important was their ability to obtain, transmit, and receive knowledge that is not generally available, and then get that information to people who can use it.[23] For multinational corporations, relational consulting allows the firm to gather best practices from one unit in one culture, to adapt those practices to operations in other cultures, and to differentiate culture-specific practices from ones that can be generalized to other operations.[24,25] It focuses on flexibility and adaptability, not on consistency for its own sake.

However, this does not mean that relational consulting does not face challenges. The most important ones stem from the opposite end of the flexibility-consistency trade-off. Maintaining flexibility is a very difficult process; it is much easier, and often much less expensive, to simply impose a one-size-fits-all package. If consultants are not careful, they can fall prey to fads and fashions. Because they have information that other members of the orga-nization do not have, it is easy (and tempting) to use that information to their advantage. Because new processes and techniques are novel and excit-ing, proposing them can enhance one's credibility.[26] In addition, once a new technique is successfully implemented in one part of an organization, there are pressures to use it elsewhere, even if it is not culturally or strategically appropriate. If relational consultants cannot resist these pressures toward knowledge isomorphism, they may sacrifice the advantages that come from the relational approach.[27]

Relational Consulting at CHips

The team's experience at CHips illustrates just how complicated organiza-tional change can be in a global organization. It was clear from the outset that the organization's management wanted to be able to exert more control over its far-flung operations, in part for functional reasons but also because centralization was a part of the organizational mindset. Eventually, they fig-ured out that what they really wanted was to have adequate information to be certain that they were making the best decisions possible, and to have high levels of efficiency. One way to achieve both goals is through a central-ized command and control structure. But, there are other ways to achieve those goals while adapting to cultural differences.

It was less clear to the team how much of a challenge it would be to manage cultural complexities during system design and implementation. Designing the new systems was difficult enough, given the realities of time and space. But, even the simplest and most concrete steps were more difficult to imple-ment than they expected them to be. Two examples were most striking. The first involved the staggered lunch and break system. In all cultures, people work primarily to support themselves and their families. But other factors, including relational ties, are also important motivators. In some settings, such as CHips' Philippine plants, those relational motivators are enacted through rituals, like lunch and break times. Once again, the importance of informal rituals is not limited to Asian cultures. One of the classic studies of work groups is Donald Roy's "Banana Time," which shows how a work group's solidarity and productivity were linked to a series of seemingly silly rituals through which they made their otherwise boring days more interest-ing.[28] There was little question that the staggered break system made the plants operate more efficiently; there also was little question that the work-ers felt that their relationships with one another were more important than increased efficiency. Existing ways of thinking and acting make it difficult for people to accept new knowledge and new techniques. In the language of relational consulting, old knowledge is sticky, making it difficult for work-ers to acquire and assimilate new knowledge.[29] In Korea, cultural deference to hierarchy both helped and hindered the team. But, the same cultural assumptions that led the staff to revere the VP and the team led them to reject the credibility of the student liaison they left behind. The difference was not a function of expertise, in terms of either his engineering skill or his knowledge of Korean language and culture, but it was a function of his formal status in the hierarchy of the culture.

As we have indicated, at this point the intervention seemed to have been a success. However, both situations needed constant monitoring. Hofstede's research found that Philippine culture values autonomy, being left alone to do one's work. The team's experience was that Korean creativity meant they started making engineering changes before the ink was dry on the blueprints. In both countries, giving local management space to adapt company systems to their individual situations seems to be necessary. But, over time both cultural tendencies threaten to recreate a situation in which headquarters will once again become out of touch with what is going on in its distant operations, and to make it progressively more difficult for each of its operations to learn from the experiences of the others. As the team recognized by the middle of the process, there needed to be a liaison in each plant, whose primary job involves communication. But, maintaining long-term contacts is the essence of relational consulting.

Review and Study Questions

1. What are the key issues associated with cross-national operations?
2. Define the two types of consulting presented in this chapter. Explain the pros and cons of each type under the environment of cross-national operations.
3. Make a list of the different situations encountered by the consulting team as they interacted with the personnel in the Philippines and Korean companies. For each situation, indicate if the communication problem is due to culture or local manufacturing requirements. Propose a solution to each problem.
4. Investigate the key differences between the Philippine and Korean cultures with respect to manufacturing environments?
5. Analyze the relationship between the American upper management and the Philippine and Korean local management. Discuss how the consultants interacted with both groups.

Notes

1. Matthew Semandini, "Toward a Theory of Knowledge Arbitrage," in *Current Trends in Management Consulting*, ed. Anthony Buono (Greenwich, CT: Information-Age Publishing, 2001), pp. 43–70.
2. Geert Hofstede, *Culture's Consequences* (London: Sage, 2001).
3. Philippe D'Iribarne, *La logique de l'honneur* (Paris: Seuil, 1989).
4. Philippe D'Iribarne, "The Honor Principle in the Bureaucratic Phenomenon," *Organization Studies* 15 (1994): 81–97.
5. Barbara Parker, *Introduction to Globalization and Business* (London: Sage, 2005).
6. Christopher Bartlett and Sumantra Ghoshal, *Managing across Borders* (Cambridge, MA: Harvard Business School Press, 1998).
7. Charles Conrad and Marshall Scott Poole, *Strategic Organizational Communication in a Global Economy* (Belmont, CA: Thomson/Wadsworth, 2005).
8. Fons Trompenaars and Charles Hampden-Turner, *Riding the Waves of Culture: Understanding Cultural Diversity in Global Business* (New York: McGraw-Hill Professional, 2001).
9. Michael Granovetter, *Getting a Job* (Cambridge, MA: Harvard University Press, 1974).
10. Trompenaars and Hampden-Turner, *Riding the Waves of Culture*.
11. Timothy Clark, *Management Consultants* (Philadelphia: Open University Press, 1995).
12. Thomas Friedman, "Imagination and Bandwidth Can Create a Global Player," *Houston Chronicle*, September 22, 2006, p. B9.
13. Erik Abrahamson, "Management Fashion," *Academy of Management Review* 21 (1996): 254–285.
14. Henri Savall, Veronique Zarde, Marc Bonnet, and Rickie Moore, "A System-Wide, Integrated Methodology for Intervening in Organizations," in *Current Trends in Management Consulting*, ed. Anthony Buono, pp. 105–126.
15. Benjamin Barber, "Some Problems in the Sociology of Professions," in *The Professions in America*, 2nd ed., ed. K. S. Lynn (Boston: Houghton Mifflin, 1965), pp. 15–34.
16. William Starbuck, "Learning by Knowledge-Intensive Firms," *Journal of Management Studies* 29 (1992): 713–740.
17. Kate Walsh, "The Role of Relational Expertise in Professional Service Delivery," in *Current Trends in Management Consulting*, ed. Anthony Buono, pp. 23–42.
18. George Hanlon, "A Shifting Profession," in *The End of the Professions?* ed. J. Broadbent, M. Dietrich, and J. Roberts (London: Routledge, 1997), pp. 123–139.
19. J. Pfeffer, *Power in Organizations* (Marshfield, MA: Pitman, 1981).
20. R. Henderson and K. Clark, "Architectural Innovation," *Administrative Science Quarterly* 35 (1990): 9–30.
21. I. Nonaka, "A Dynamic Theory of Organizational Knowledge Creation," *Organization Science* 5 (1994): 14–37.
22. I. Hargadon and Robert Sutton, "Technological Brokering and Innovation in a Product Development Firm," *Administrative Science Quarterly* 42 (1997): 716–750.

23. Robert Burt, "The Contingent Value of Social Capital," *Administrative Science Quarterly* 42 (1997): 339–365.
24. J. J. Kao, *Jamming: The Art and Discipline of Creativity* (New York: Harper Business, 1996).
25. Leonard Kraar, "The Overseas Chinese," *Fortune* 130, no. 9 (1994): 91–114.
26. E. Abrahamson, "Management Fads and Fashions: The Diffusion and Rejection of Innovations," *Academy of Management Review* 16 (1991): 586–612.
27. Paul Dimaggio and William Powell, "The Iron Cage Revisited: Institutional Isomorphism and Collective Rationality in Organizational Fields," *American Sociological Review* 48 (1983): 147–160.
28. Donald Roy, "'Banana Time': Job Satisfaction and Informal Interaction," *Human Organization* 18 (1950): 158–168.
29. George Szulanski, "Exploring Internal Stickiness: Impediments to the Transfer of Best Practice with the Firm," *Strategic Management Journal* 17 (1997): 27–43.

11

Bucket Brigades Work, but Why?

In Chapter 1 we explained that engineering as a profession has entered an exciting and challenging new era. Organizations now operate globally, not locally or even nationally. Global competition, complicated organizational structures with fluid and confusing lines of authority, operations that are located thousands of miles from one another, and work teams that include members from diverse cultural backgrounds, whose first languages are very different from one another, and who are accustomed to very different ways of doing things, have become the normal way of doing business. However, within all of this complexity there are some recurring patterns. All of the situations that today's engineers face are influenced to some degree by the global orientation of their organizations and the organizations with which they have contact—their global mindset and the structures through which that mindset in implemented. All organizations must establish and maintain successful strategies for obtaining necessary resources, understanding and responding to market pressures, and developing efficient operations and supply chains. Finally, global organizations must adapt to the cultural, economic, and political realities within which they operate.

The challenges that today's engineers face result from the need to be aware of all of these interdependencies, and to be able to determine which ones are most important in particular cases. In the MEPO and USAHP cases, cultural factors were crucial, particularly in terms of how they related to the companies' global orientations. Even an especially insensitive consulting engineer did little damage to MEPO's operations because its employees on both sides of the Atlantic had developed effective working relationships during years of direct contact. Conversely, even a well-designed global orientation and implementation strategy initially failed at USAHP because the company underestimated the impact of cultural and subcultural factors. SmartDrill's management made wise relocation decisions both because it took cultural, operational, and economic considerations into account, and because it refused to take anything for granted. Although cultural factors were indirectly relevant at HOCH, the core of its problems involved operational factors—the inevitable complications of outsourcing, collaboration among multiple companies, and cross-organization power relationships. In AAC's international supply chain problem, differences across cultures were relatively unimportant. Internal aspects of the organization and its decision making, including management's commitment to a narrow definition of the problem it faced, were paramount.

In this chapter we present our final case study. Like Chapter 6, it involves an AAC plant. Unlike the earlier cases, we will not conclude the case with our interpretation of the events that took place. Instead, we will provide three plausible explanations and ask you to develop a final interpretation. By concluding the book in this way, we hope to give you an opportunity to apply the concepts that we have discussed, and thus to better understand how they apply to real-world situations. In order to disguise the identity of the product in the real case study, in this section the product produced in AAC will be assumed to be a leather car seat assembly.

The Facts of the Case

The manager of AAC's interior seating systems plant in Azteca, Mexico, held a regular monthly meeting with his department managers. He announced that he was going to bring an American consulting group to the plant to take a look at some of their assembly lines and make change recommendations to improve throughput. He asked that the material and operations managers be in charge of assisting the consultants with anything they needed. The consultants were going to start with the leather car seat assembly process, since this was one of their least productive lines.

The following week, the consultants came to the plant and met with the plant, material, and operations managers. After taking a tour of the facilities and receiving the information that the plant managers had put together for them, the consultants told them that they would be there for the next couple of weeks collecting information about the leather car seat assembly line that they had discussed. One of the consultants, Luis Robles, informed them that within 3 weeks they would present their plan of action for improvements.

Luis went to the assembly line to study the existing process. He saw that all of the operations involved some manual assembly, and all the operators were sitting on tall stools. Once an operator was finished with an operation, he or she tossed the assembly to the next station. Since different amounts of time were required to complete the tasks performed at each station, some assembly accumulated in front of some workstations. Since there was no restriction as to how many parts could be accumulated in between stations, some of the piles got rather large. From previous experience, Luis realized that there were problems involving work in process (WIP).

In theory, assembly lines are designed to be balanced so that there is little variability in the time required to complete each task at each station. Items being assembled move smoothly from one step to the next. More complex or more time-consuming tasks are broken into work elements, usually through the use of time-motion studies, and an appropriate number of work elements are assigned to an assembly station. Typically, each assembly station is operated by a different worker.

Even well-designed assembly lines can become unbalanced. For instance, unforeseen breakdowns will occur affecting different assembly stations in the line at different random times. Once the broken down station is repaired, it will restart with work that has accumulated during the down period. Operators are also an important source of variability. If for some reason some of the workers are less motivated than the others, bottlenecks can occur at their position on the lines. Similarly, workers who are better trained and have more experience will tend to work faster than those with less effective training or little experience. Consequently, if the organization experiences high levels of turnover,[1] lines may become unbalanced because neither temporary replacements for absent workers nor new employees hired to replace those who quit are likely to work as quickly or effectively as experienced workers. If turnover is especially high for specific stations on the assembly line, it can quickly get out of balance. Bottlenecks quickly develop and the line has to be slowed down until the bottlenecks are cleared. Similar assembly line imbalances will be caused by absenteeism.

Analyzing the Causes of the Problem

After making his preliminary observations, Luis decided to conduct a line balancing evaluation as a starting point in trying to improve the efficiency of the production line. In order to begin his analysis, he gathered some information about the car seat assembly and developed a precedence diagram (Table 11.1) of each step in the procedure, with the time that it takes to complete each. He wanted to study how the current workstations were laid out to see if he could come up with an improved balanced layout of the assembly process.

From the data it can be observed that technically the bottleneck is station S-360 because it requires the highest amount of work (measured in time). However, stations where sewing leather is required can be a problem, as the sewing is done manually due to absenteeism and medical leaves. This operation is very stressful on the operators' fingers and hands.

Once Luis completed the line balancing analysis, he thought he had a pretty good idea about where to begin recommending changes. If plant management would address the variability of assembly times among assembly stations, they would have a more balanced line, and that would greatly reduce the amount of WIP. He thought, however, that before he went any further, he should show his results and analysis to his boss, Pablo Castillo. Pablo decided to have a meeting to discuss the information. Pablo began the meeting by telling Luis: "After looking at your data and analysis, I agree that their assembly time variability is definitely a problem to address. However, don't be so sure that this is the only cause of their problems. I want you to go back and do some more analysis. Some of the reports about the assembly line mentioned that they had high turnover rates; that is very common in

TABLE 11.1

Precedence Table of Car Seat Assembly

Station	Element	Preceding Elements	Time at Station in Minutes
S-005	Removing paper from panels	—	0.94
S-120	Sew hems of central panel	S-005	0.56
S-130	Adjusting central panel	S-120	0.54
S-310	Join first lateral panel	S-130	0.76
S-340	Reinforcement of first lateral panel	S-310	0.74
S-345	Auxiliary seam on the central panel	S-340	0.69
S-010	Mark restraints	—	0.51
S-015	Sew reinforcement on restraint panel	S-010	0.59
S-040	Closing seam on restraint panel	S-015	0.60
S-050	Sew seam around edges	S-040	0.87
S-350	Join restraint subassembly	S-345, S-050	0.82
S-360	Join second lateral panel	S-350	1.19
S-370	Reinforce second lateral panel	S-360	0.98
S-430	Seat assembly inspection	S-370	1.01
S-480	Place metallic restraint and folding	S-430	0.75
S-500	Seat assembly, pressing, shrinking, and sealing of film	S-480	0.56
S-540	Place seat assembly wrap, labeling and packaging	S-500	0.96

maquiladoras in Mexico. Think about how this is affecting their assembly line productivity."

Luis responded, "Good point. Not only are the assembly times unbalanced, but it doesn't help when there are new workers coming in all the time starting from scratch."

"Yes, this problem is a little more complicated," said Pablo.

"Why is it that the employee turnover rates are so high?" asked Luis.

"Well, that is a good question ..." Pablo responded. "*Maquiladora* plants in Mexico are known to experience high turnover rates for many reasons. In some border cities low wages and agricultural migration during harvest seasons contribute to that problem. Also, if you have a small *maquiladora*, it will offer lower wages and no continuing education for its workers so that they can gain skills. On the other hand, if you are a larger, well-established company, then you offer all those things and you can manage to keep your turnover rates down. For most border cities, their turnover rates range from 10 to 20%; when you compare that to the average U.S. turnover rate of 3%, that makes a big difference!"[2]

"Look here, Luis, I searched some of the numbers for the city of Azteca," he said pointing to some papers, "and since it is approximately 230 miles from the border, the turnover rates are a little lower. The average is 4.5% per month—still higher than the U.S. average."

"Yeah, but in the months of January and March, when there are more jobs in agriculture, the turnover rates get up to 7%!" Luis said with surprise. He continued, "That would really hurt AAC with developing a consistent production rate and also affect our training budgets. Thank you for the information, I will take this into consideration."

Effects of Turnover

Labor turnover impacts throughput of a serial assembly line in two ways: the length of time an operator stays at his job (tenure) and how quickly an operator learns his tasks (learning curve). These factors are considered uncontrollable because they depend on the motivations and skills of each individual worker. The tenure of a worker at AAC is illustrated in Figure 11.1, which depicts how many days (tenure) versus the number of employees (frequency) that remain in their position for that period of time. As Figure 11.1 indicates, 62% of AAC's assembly line workers quit within 200 days of being hired. Although this figure may seem high, it is consistent with historical experience in the industry. For example, after the Ford Motor Company shifted to an assembly line system in 1913, it had to hire 963 workers in order to have 100 left at the end of the year, a total of 50,000 workers in order to retain the 14,000 needed to run the Highland Park plant. Ford responded by increasing the wages paid to the workers to the unheard level of $5 per day, but the problems continued until the Great Depression effectively eliminated job options for Ford's workers. But, low job satisfaction and high turnover

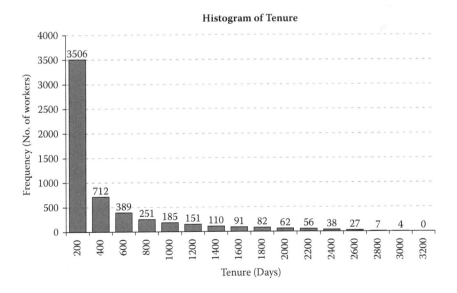

FIGURE 11.1
Histogram of labor tenure at AAC.

seem to be inherent in assembly line work. Fifty-five years after opening the Highland Park plant, turnover was still a problem for the Ford Motor Company. In 1969 almost half of Ford's assembly line workers quit within 90 days of being hired. Eight percent of Ford's workers quit every month, so the company had to hire an entire new workforce every year.[3]

In order to analyze the assembly time variability caused by the introduction of inexperienced workers into the assembly line, the consultants applied a technique called the learning curve (LC) theory. This theory specifies how the time it takes for a person to complete a task reduces with practice (i.e., the number of times the person has repeated the task). This analysis yields a function, called the LC, that represents the time to complete a task as a dependent variable, and the number of parts completed as the independent variable. The data were collected from AAC from the traditional assembly line (see the appendix).

Dynamic Work Allocation Methods (Bucket Brigades) to the Rescue

The solution that the consultants needed had to address the problems of both high variability in production time and high employee turnover rates. In addition to the inherent variability between assembly stations, turnover is especially harmful to assembly lines since less experienced workers will not be able to perform their tasks as quickly as more experienced ones, which will generate blockage and starvation in successive stations, which in turn causes a reduction in the line's throughput. This is especially true when dealing with tasks that require significant skill since they may have a significant learning curve. Working with leather is challenging because it is a natural material and the raw material can vary from batch to batch, requiring a skillful operator that can produce the same good finishes despite the natural variations of the material. There are several work allocation strategies for serial assembly lines. Particularly, there are some that address the high turnover rates called dynamic work allocation (DWA) techniques to design serial assembly lines. Instead of requiring organizations to engage in complicated systems for keeping lines balanced, DWA methods utilize *unbalanced* line designs. These unbalanced lines are used to absorb the process variability that results from the turnover problem. Unbalanced systems are not all that complicated in themselves. In fact, in a famous case involving the workers at a Chevrolet Vega plant in Lordstown, Ohio, a system was created in which workers would double up, doing both their job and the job of the next worker on the line. While the first worker doubled up, the second rested. Then they switched roles. Although the system increased job satisfaction and reduced absenteeism and turnover, GM punished them for creating the system because it had not been approved by management. But, the example shows how valuable unbalanced systems can be, and how easy they are to implement in some cases.[4]

However, the problem at the Azteca plant was more complicated because the consultants had to address both high turnover and high WIP levels, which

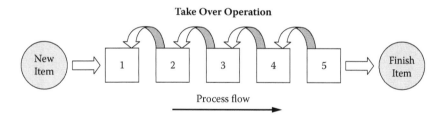

FIGURE 11.2
Bucket brigade method.

seemed to be related to additional factors. In order to do so, the consultants decided they needed a technique that could be implemented as a continuous assembly line where each assembly station works on a single assembly at a time. All of these requirements left them with a couple of choices, and they opted for a bucket brigade system.

Bucket brigade (BB) is a DWA method to organize assembly line operations in such a way that the line is self-balancing. They found this method to be advantageous because there would be no need for costly time and motion studies or cumbersome line balancing practices. In addition, this method would reduce the level of supervision required of management. In a sense, BB recaptures the primary advantage of using assembly lines in automobile manufacturing. Before shifting to assembly lines, each supervisor at Ford Motor Company could only effectively oversee five to seven men. Afterwards, the span of control increased to fifty-nine, allowing Ford to massively reduce the number of highly paid supervisors that it had to hire. In addition, the machinery itself "hurried up the slower men."[5] In BBs workers start to function as a self-organizing production system that reaches its optimal structure spontaneously. Work elements are not assigned specifically to an operator or workstation; instead, they are dynamically assigned based on the workload, relative speed, and experience of the operators. Because of BB's ability to self-balance and absorb variability in the system, it minimizes the impact of turnover.

The next step for the consultants was to develop the process for the BB method. The operators are arranged from slowest to fastest, as in a relay race. This way the faster operators down the line make up for any lag caused by the previous operator. Operators work on the part until the next operator interrupts them, as can be illustrated in Figure 11.2. This process is known as preemption.

An operator utilizing this method would have to remember the forward and backward rules:

Forward:

- Each operator should work on the seat assembly until the next operator down the line takes the seat assembly.

- Each operator should move further down the line as possible, completing as many operations as possible.
- When the next operator down the line requests the part, the downstream operator should relinquish the seat assembly and follow the backward rules.

Backward:

- After relinquishing the part, each operator should walk upstream the line and take over the seat assembly of the first operator that they find on the way back.
- After the downstream operator has taken over the seat assembly of the first operator of the line, he should introduce a new seat assembly into the line.

These rules are designed to encourage the operators to work on different operations when trying to complete as many operations as possible on the seat assembly. If the number of units assembled increases, then everyone in the line will benefit, especially if they are paid under a piece-rate system. Notice that if operators are not allowed to move between stations, the number of pieces produced by the assembly line will be determined by the slowest station, and every operator assigned to the line will be limited to the same low production rate. By comparing the production rates and efficiencies of a different number of operators, they found the number of operators they needed applying a technique called discrete event simulation; the technique involves mimicking the real system operation in the computer and simulating a variety of scenarios until a good line design is encountered. Workstation assignment (range/region) and assigning work elements to workstations were also done through the simulation study.

In the design of a BB assembly, workers are assigned to a region of workstations, and not to any particular station. One way to arrange the region is to divide the workstations among the workers. The regions should overlap (Figure 11.3). This method does not require an accurate measure of task times. However, BBs emphasize that the operators work as a team in order to function at their best. Ironically, this team structure is very much like the one that existed at Ford Motor Company before it switched to assembly lines. Similar systems were developed by Volvo during the 1980s, and by the line workers at the ill-fated General Motors Lordstown plant.[6]

Proposal

The consultants utilized the results of their simulation of both the traditional assembly line and the BB system in a presentation for the managers at AAC. The consultants knew that this was a major change from AAC's current methods and expected a substantial degree of resistance. Not only would the changes affect the managers, but the operators in the production line

Traditional Line Balancing Method

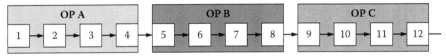

Bucket Brigade Method

FIGURE 11.3
Diagram of traditional vs. BB methods.

would have to become accustomed to these new ideas. For the workers, the greatest change would be requiring them to remain standing, instead of sitting. Standing is required under BBs to allow an operator to move from station to station. Also, operators would have to get accustomed to the takeover operation, where the downstream operators had to interrupt the operation and take over from there.

The managers were concerned about some of the changes. The material manager asked: "What about the ergonomic implications now that you propose that the workers are standing instead of sitting? They are going to get tired after a full shift of 8 hours!"

Luis answered, "Actually there are more advantages to standing, because they will not just stand there. When the workers are just standing in a static position, then you can be concerned about ergonomic problems. With this method they have to walk back and forth between stations, which will keep them constantly active. Since they will always have something to do, they will have no reason to be standing in static position."

"What about the expected production yield of the workers, will they be required to turn out more volume?" asked the operations manager.

"Well, we expect that your productivity will increase, of course. In our simulation model we took into account that the pace of the workers will be affected by their learning curve. At first it will take some time for them to get used to it," said Luis.

"The results that you showed us today sound really promising, and I really like the one-piece flow idea. You say these are the latest trends in assembly line design, right? Well, it sounds great. Let's do it," concluded the plant manager.

Implementation

On the first day of implementation the consultants held staff and operator training of the three production shifts. Since the BB method represented a drastic change, there were several hurdles that the consultants encountered and had to be overcome to make their implementation successful.

One of the first hurdles was the operators' resistance to the new methods. They responded to the idea by saying "no way," "crazy." They did not like that they had to change from a sitting position to a walk/stand position. The operators were completely against having their chairs taken away. Luis was astonished at how much they were resisting, although he also realized that most *maquiladora* workers hold more than one job. They work long hours, up to 16 hours a day. People on the second and third shifts usually were tired by the time they got to work. With respect to American standards, these workers were also underpaid. The operators at this assembly line liked their job because they got to sit down and work, and even sleep when bottlenecks developed. They believed that the traditional assembly line was comfortable and easier.

A second hurdle was the managers' opposition. Before Luis and the other consultants could do anything about it, some operators where protesting against this new method to their cell leaders. From the cell leaders this would be communicated up the chain of command to the operations and human resources manager. The HR manager at this plant had sufficient amount of interaction with the operators of the production lines to have become involved in the project, since they are responsible for disciplinary actions or firing of employees. Consequently, he had a lot of influence over the workers. The material manager was behind the new ideas that the consultants had proposed, but he was the only supporter among the mid-level managers. The HR manager and the production manager had political motivations for opposing the new system—they were concerned that the material manager would be able to take credit for the new system if it succeeded. Even though the HR and the production manager were at the same level in the organizational hierarchy as the material manager, the HR manager had more political power in the company. He was tightly linked to the company's HR director, who was above the plant manager in the company's chain of command. Much like the USAHP case study described in Chapter 5, when the managers did not show full support to this new project, the lack of enthusiasm trickled down to the rest of the employees, resulting in little cooperation from cell leaders and operators. In addition, like in the USAHP case, plant management was not willing to offer any financial incentives to encourage the workers to support the new method. A great deal of change was demanded from the operators, change that had obvious costs for them with no motivation to accept. Their resistance led to an actual reduction in assembly speed and productivity during the first week of implementation.

The third hurdle that implementation faced was related to supervision. The new method was not properly sold to the cell leaders of the first and third shifts. There was a lack of trust in the advantages offered by the new method. This led to their implementing the system in a way that undermined its operation. For instance, the order of the operators, which should always be from slowest to fastest, was overlooked. Faster operators were assigned to the beginning of the line, virtually converting the pull system into a push system. The one-piece flow was also frequently overlooked, since it was common to find several seat assemblies in between workstations, and this caused the operators to remain static in a workstation for long periods of time, jeopardizing the ergonomic advantage of the stand/walk characteristic of the BB method. Luis realized that it was imperative to have everyone's cooperation to make this new method work. The operators of the BB method required discipline and teamwork, and they were not used to it. They would have to learn the new rules and practice them very strictly for the method to be successful.

The fourth hurdle in implementation was the high frequency of machine breakdowns. When following a one-piece flow methodology, any machine breakdown impacts the line more drastically than when buffers are present, since buffers absorb the variation caused by breakdowns. When a duplicated machine is down, the impact is not as strong as when a unique machine is down, in which case the line stops completely, and after repair, the line needs time to stabilize. The impact that machine breakdown has on team morale is also an important factor that is difficult to quantify but plays an essential role in line performance.

Hurdles Overcome

In order to overcome the hurdles, Luis decided to spend more time talking to the workers, taking suggestions from them about how the process could be improved. He let the workers feel as if they were creating the process themselves, and that made the workers feel important and responsible for its success. To work with the Mexican people, Luis realized he needed to gain their trust. One day Luis spent an entire 24-hour day with every shift, showing them how to do the proper procedures and making sure that production went according to plan. His level of involvement impressed the operators, and a good relationship was built between them.

Once Luis realized why the operators were so resistant to the no chair aspect of the new process, he understood better how to approach this problem. Luis tried to address their concerns by adding ergonomic mats on the floor of the production line that were designed to reduce fatigue. He also tried to win over the workers with the idea that the stand/walk position would increase their fitness and make them stronger, an important value in Mexico's highly masculine culture. Luis also explained about the physical harms that could occur when sitting for too long. He tried to convince them that the new method was crucial for the plant's survival and for the company to remain competitive.

In relation to opposition from the supervisors, Luis realized that he had to make the right connections with the right people. He realized the importance of a female group leader named Yolanda, who had been promoted from the bottom of the hierarchy and was extremely tough. She could inflict such strong verbal injury that people would cry because of her. Yolanda was in charge of managing the production cell. Her responsibilities included quality assurance, keeping track of production on an hourly basis, reporting daily production quotas, and assigning people to positions on a daily basis. Workers relied on her to get their jobs for each day (some jobs are harder than others). She had the hidden power both up and down the line of command. Her influence was beyond her scope of responsibilities. She was perceived by upper management to be the whip to get work done by the end of the day. If she didn't buy into the ideas, nothing was going to happen. Fortunately, Luis was able to relate well with her and convinced her to help him implement the new method.[7]

Regardless of the hurdles and resistance by operators, cell leaders, shift managers, and managers, the line was implemented. In contrast with the traditional line, the operators in the BB line are always busy. Due to the one-piece flow methodology, the operators always know what to do next. If the next operator takes away the part from them, they then follow the backward rules and preempt the previous operator. Several operators have stated that they feel as if the shift goes by faster working under this production system, since they always have something to do.

Results

After Luis dealt with worker resistance, the worker's learning curve with the new system started to climb. Eventually, the implementation resulted in high productivity rates. The production goal for all shifts is based on a 70 parts per hour rate. That rate corresponds to the maximum rate assigned to these types of lines at AAC. The maximum production rate of all shifts was 64.75 parts per hour, achieved by the second shift. The first and third shifts' maximum production was 63.5 and 64.2 parts per hour, respectively.

In the past when AAC was faced with the production requirement of 70 parts per hour, it was common to use more than twenty-one operators to meet the production quota. The number of operators in the first and second shifts always fluctuated between fifteen and nineteen. On average, the number of operators used in the BB production system is two less than in the preexisting method. From the outset, the highest-performing line was the second shift, the one that had a supportive supervisor from the beginning. Under Yolanda's influence, that line reached the production goal of 70 parts per hour during 3 consecutive days.

A Turnover Interpretation

This is the interpretation that the consulting team assumed was accurate throughout their intervention. However, there are a number of gaps in the evidence that was made available to the team. For example, the team had no data that broke turnover rates down by tasks performed. AAC's leadership assumed that the turnover rate was highest for the tasks that led to bottle-necks in the original assembly line. Those workers should have experienced a great deal of stress as BBs piled up and the other workers waited when the line was stopped for the workers at the bottlenecks to catch up. Management also assumed that turnover was most damaging at these points on the line because those employees had complicated tasks, which involved lon-ger learning times than the other activities. As a result, the negative effects of high turnover on those tasks would be exaggerated by the longer time needed to train replacement workers. But, they also lacked direct evidence on the learning curves for each task. In addition, there is some anecdotal evi-dence indicating that many workers welcomed the bottlenecks. It gave them a chance to rest, or even sleep. Given this, why would they hassle the work-ers at the bottlenecks? Of course, just knowing that you were responsible for stopping the line might be stressful in itself, but at this point there is no direct evidence of that. Determining why people leave their jobs is difficult unless an organization has some systematic form of exit interview, one that employees participate in honestly. When employees just fail to show up at work, it is impossible to know exactly why they made that decision. To com-plicate the matter even more, while it is clear that productivity is up under the BB system, it is not clear that turnover has been reduced. The consulting team left once the implementation problems seemed to be ironed out, and management has been so happy with the improved productivity that they have not been paying much attention to turnover problems. Consequently, it is difficult to pin the problems of the balanced line on high turnover, and it is equally difficult to attribute the success of BBs to reduced turnover. Since AAC's other plants vary widely in terms of the extent to which they have turnover problems, management is hesitant to make major changes in those plants, especially those that have low turnover rates.

A Motivational Interpretation

Traditional assembly line work is boring, repetitive, tedious, and exhausting in even the best situations. It requires relatively low levels of skill, demands that workers show absolutely no initiative or creativity, and provides little opportunity for workers to feel pride in a job well done. For example, in 1910, before Ford shifted to assembly lines, the federal government classi-fied 75% of all the jobs in car manufacturing. By 1924 the figure had fallen to less than 10%. This change was especially humiliating for the best of Ford's workers—one worker told Peterson about his father's experience with the transition: "It was a sad thing for a child to see his father's job, the source of income, the security of the family, taken away, as it were, piece by piece, to have your father reduced to the status of other men in the neighborhood who were just plain ditch-diggers."[8] BBs regain some of the opportunity for pride that disappears in balanced assembly lines. In addition, coordinat-ing the workers is a more complex task, one that must be performed by the workers themselves. As a result, they talk to one another more. Luis started spending much more time on the production floor communicating with the line workers, just as Ford's supervisors did before the introduction of the assembly line. All of the key elements of job satisfaction—satisfaction with supervision, feelings of achievement, relationships with co-workers—should be greater with BBs. If the increased productivity observed in Azteca is the result of a more stimulating and motivating work situation, AAC should be able to implement it successfully in all of its other plants, regardless of their turnover record. But, the company has no job satisfaction data available, either for the years before the change or since.

Review and Study Questions

1. Classify the facts of the case study using the GEM model.
2. Which is the predominant layer in this case?
3. Describe the interactions between facts in different layers of the model. Do you think that nontechnical factors were significant in the outcome of the project?
4. Two possible interpretations of the results were presented. Which one do you think has more validity? Do you have another interpretation?
5. Assume you have to implement the same assembly line in the United States and Mexico. What differences do you expect to observe during their implementation? What modifications would you suggest to tailor the original design for each country?

Appendix

Learning Curve Data (Time in Seconds per Part over Time)

	Sample Number									
Day	1	2	3	4	5	6	7	8	9	10
1	91	94	90	103	85	96	100	91	97	95
2	82	73	74	72	81	70	73	77	73	67
3	63	71	77	70	70	65	69	69	76	71
4	64	69	64	69	69	63	73	65	60	62
5	61	71	59	63	65	63	58	60	60	66
6	63	61	57	63	59	60	59	64	60	61
7	71	59	59	62	62	64	56	58	59	
8	64	61	58	63	66	65	67	63	63	67
9	62	59	55	60	56	60	59	58	59	60
10	50	56	53	56	53	49	52	52	60	53
11	60	52	51	54	51	52	58	51	53	52
12	49	62	54	52	51	52	52	54	48	53
13	51	51	59	46	54	55	45	50	51	54
14	54	56	51	56	55	52	52	55	52	50
15	65	61	54	61	55	53	58	53	51	51
16	54	48	51	51	52	48	53	43	50	54
17	49	51	50	45	53	45	42	48	45	47
18	46	49	48	47	45	47	51	46	43	52
19	45	44	44	44	44	46	45	40	46	44
20	46	45	47	46	45	46	46	43	46	46
21	41	44	46	44	47	41	44	43	44	44
22	42	44	44	45	42	42	42	46	45	44
23	45	45	42	46	43	46	45	43	44	45
24	45	43	43	40	41	41	40	41	41	39
25	45	41	44	42	47	44	41	40	38	43
26	45	44	43	44	41	43	43	42	43	41
27	44	41	43	45	45	42	43	42	44	41
28	44	44	44	41	42	40	41	40	40	44
29	41	41	39	45	42	45	43	41	45	39
30	45	42	40	43	46	40	41	43	46	42
31	43	45	39	39	45	42	42	43	42	42

Notes

1. Turnover rate is the frequency at which workers separate from the company in a specific time period.
2. Information taken from http://www.teamnafta.com/ and http://data.bls.gov/cgi-bin/surveymost, 2003.
3. John Eldridge, Peter Cressney, and John Macinnes, *Industrial Sociology and Economic Crisis* (New York: St. Martin's Press, 1991).
4. "The Lordstown Auto Workers," in *Life in Organizations*, ed. R. M. Kanter and B. Stein (New York: Basic Books, 1979).
5. Henry Ford, cited in Richard Edwards, *Contested Terrain: The Transformation of the Workplace in the Twentieth Century* (New York: Basic Books, 1979), p. 118. Also see Henry Braverman, *Labor and Monopoly Capital: The Degradation of Work in the Twentieth Century* (New York: Monthly Review Press, 1974).
6. Christian Berggren, *Alternatives to Lean Production* (Ithaca, NY: ILR Press, 1992).
7. For an analysis of the complex political alliances that develop in organizations, see Henry Mintzberg, *Power in and around Organizations* (Englewood Cliffs, NJ: Prentice Hall, 1983).
8. Joyce Shaw Peterson, *American Automobile Workers, 1900–1933* (Albany, NY: SUNY, 1987), p. 39.

Index

Printed and bound by CPI Group (UK) Ltd, Croydon, CR0 4YY

18/10/2024

01776244-0005